DIRECTORY OF

NATIONAL COMMUNITY VOLUNTEER FIRE PREVENTION PROGRAM

PARTNERSHIPS
AGAINST
FIRE

COMMUNITY-BASED FIRE PREVENTION EDUCATION INITIATIVES

1984-1992

Revised February 1993

Disclaimer

The development of this directory was supported under funding by the Federal Emergency Management Agency's United States Fire Administration, under contract No. EMW-89-C-3059. Any points of view or opinions expressed in this document are the results of work performed by the National Criminal Justice Association and project staff and do not necessarily reflect the official position or policies of the Federal Emergency Management Agency.

TABLE OF CONTENTS

PREFACE

The National Criminal Justice- Association (NCJA) is pleased to provide this publication, *Directory of National Community Volunteer Fire Prevention Program Community Based Fire Prevention Education Initiatives: 1984-1992.* This revised Directory updates the original version published in 1990 and is intended for public safety officials and private sector organizations wishing to learn more about program development, resource-raising, and specific NCVFPP activities.

The Directory contains information on programs and program contacts of all the local fire prevention programs that have continued to operate since completing their participation in the NCVFPP and includes information about programs initiated since the original *Directory's* publication. Appendix 1 provides information about the remaining programs developed and implemented under the NCVFPP, but not continued once federal funding was completed, or for which no updated information was provided.

We would like to extend our gratitude to all our state and local National Community Volunteer Prevention Program (NCVFPP) colleagues, public safety officials, fire service personnel, social service agencies, educators, and private sector organizations who have been so instrumental in implementing and maintaining the NCVFPP. It is our hope that the *Directory* will enable its users to build upon established approaches to community-level fire prevention and to provide guidance in developing, implementing, and assessing the effectiveness of various prevention programs.

INTRODUCTION

The *Directory of National Community Volunteer Fire Prevention program CommunityBased Fire Prevention Education Initiatives* is a principal product of the National Community Volunteer Fire Prevention Program (NCVFPP), administered by the National Criminal Justice Association under the auspices of the Federal Emergency Management Agency's United States Fire Administration (USFA). The NCVFPP is a community-based effort designed to increase the scope and effectiveness of local fire prevention and education programs.

The *Directory* provides general information about the 31 local programs developed and implemented under the first phase of the NCVFPP that were updated through a follow-up effort and the two programs developed and implemented for visually- and orthopedically-impaired individuals under the second phase of the program. Appendix 1 provides information about the remaining programs that were developed and implemented under the first phase of the NCVFPP. The publication is intended to serve as a resource for states and for other organizations and individuals interested in developing or expanding local fire and burn prevention and education programs.

The primary purpose of the NCVFPP is to achieve a significant and continuous reduction of deaths, injuries, and property losses due to fires in the United States. To achieve that goal, the NCVFPP seeks to support fire prevention efforts of community-level volunteer programs by coordinating the resources of federal, state, and local fire services with the resources of the private sector,

Over the years, the NCVFPP has supported the development and implementation of programs addressing such fire safety issues as fire and burn prevention in the home, identification and reduction of common fire hazards, use of fire extinguishers and smoke detectors, proper use of home heating devices, and fire survival and escape. Programs have addressed the general public and a wide range of special audiences, including children, adolescents, young parents, the elderly, and the handicapped.

Background

The first phase of the NCVFPP began in fiscal year 1984 with the participation of 10 states that, together, represented approximately 40 percent of the total U. S. population. Additional states were brought into the program under the first phase of the program through fiscal year 1989. By the end of fiscal year 1989, a total of 30 states and the District of Columbia had participated in the program.

Under the first phase of the NCVFPP, a state point-of-contact acted as coordinator of NCVFPP activities in the state and served as liaison between the NCJA and local grant recipients. A state committee served as a medium for bringing together state and local government, fire services, private sector representatives, and community volunteer organizations to identify specific fire control needs in the state and develop strategies responsive to those needs. Local organizations developed and implemented locally-targeted programs designed to educate citizens about fire problems specific to their communities and to train those citizens in fire prevention and survival. The USFA, the NCJA, and state points-of-contact teamed up to provide participating local organizations with funding and such technical assistance as training in formulation of fire prevention strategies, resource-raising, and management of volunteers.

Under the NCVFPP's first phase, funds were awarded through cooperative agreements to states and local volunteer groups to foster development and implementation of fire prevention and control programs tailored to particular state and local needs. In accordance with funding guidelines effective during the program's first phase, state and local participation in the NCVFPP followed a three-year funding cycle designed to provide local organizations with seed money to establish innovative and effective programs while providing for a gradual shift from federal funding to outside sources of funding and support in order to allow those programs to continue after the termination of USFA assistance. During a state's first year of participation in the NCVFPP, the state and local groups within that state received USFA funding to cover 100 percent of expenses for NCVFPP activities. During the second year of participation, a state continued to receive 100 percent funding, while local groups received 75 percent of the amount of first-year funding and provided the remaining 25 percent in the form of matching funds, materials, or services. During the third year of participation in the NCVFPP, a state continued to receive 100 percent funding, but the USFA did not provide funding to the local programs. A total of 106 local projects were funded under the first phase of the program.

In 1989 USFA officials initiated the second phase of the program, partly in response to budget constraints and partly to capitalize on experience and models already developed under the program. The revised program reflected the USFA's commitment to bring into the NCVFPP the 20 states that had not yet participated in the program. Unlike the first phase of the program, in which states and local organizations received funds, the revised program provided funds only to states, with which the states helped local organizations develop fire prevention programs in their communities without federal funding assistance. The NCJA provided technical assistance and information to states about programs and resource-raising techniques previously developed under the NCVFPP that served as models for new community efforts. Nineteen of the remaining 20 states participated in the second phase of the program. In addition, two grants were awarded to local community volunteer organizations to develop and implement pilot programs for visually impaired and orthopedically, or movement, impaired individuals. Through fEeal year 1992, a total of 49 states and the District of Columbia had participated in the first and second phases of the program, and a total of 108 local projects had been funded.

In the fall of 1992, the USFA initiated the third phase of the NCVFPP. Seeking to sustain the NCVFPP's impact in the states, the USFA developed and implemented a national program of training and technical assistance (NT & TA). Through the NT & TA program the USFA hopes to encourage states to invest financial and other resources in helping community organizations to develop and implement long range strategies to sustain and enhance the level and quality of tire prevention services. Services provided under the NT & TA program include regional, multi-state instructors training workshops; topic-specific workshops on prevention strategies development, fund-raising, and community coalition building; and on-site general technical assistance to provide consultation on fire prevention strategies development and community coalition building.

Use of the Directory

All of the local fire prevention programs initiated under the NCVFPP funding have completed their participation in the program. In the fall of 1992, a comprehensive follow-up effort was undertaken to determine the status of local fire prevention programs funded under the NCVFPP. Some programs have continued to operate after leaving the NCVFPP by modifying their activities or developing other sources of support, and in most cases where local programs have ended, the organizations that sponsored those programs remain in existence.

Both the directory and appendix 1 contain program profiles, ordered alphabetically by state and name of sponsoring organization, that provide brief descriptions of audiences targeted, problems addressed, and activities undertaken by the programs. In addition, each program profile lists the original contact person and address for the program, as well as the most recent contact person and address, if changes have occurred. These contact persons may prove a valuable resource to directory users wishing to learn more about program development, resource-raising, and specific activities under past programs.

To assist users in identifying those programs targeted toward tire prevention issues of most concern to their community, two directory appendices, appendix 2 and appendix 3, provide cross-references to programs classified by target audience (preschool-aged youth, elementary school-aged youth, secondary school-aged youth, adults, senior adults, the general population, and special populations) and by program category (school-based, residential, commercial, population-specific, and special focus tire prevention programs). Appendix 4 lists by program category those programs that developed original educational and informational materials, such as curriculum plans, brochures, video productions, and public service announcements, to complement program activities. Each appendix program listing indicates the number of the directory page on which that program's profie appears; in turn, each profile contains codes indicating that program's classification. Many of the programs fall into two or more categories because they were comprehensive in nature, seeking to address several fire issues and target multiple audience groups.

**Key to Classification Codes for
Directory Cross-Referencing**

Targeted Audience

    PS   Preschool (ages 2-5)
    ES   Primary/Elementary School (ages 6-12)
    SS   Secondary School (ages 13-20)
    A   Adult (Ages 21-64)
    SA   Senior Adult (ages 65 +)
    GP   General Population (i.e., no specific age group)
    SP   Special Population (e.g., based on disability, cultural diversity, language other than English)

Program Category

SB   School-based Fire Prevention Programs

    ES   Elementary and Secondary School
    PS   Preschool

RB   Residential-based Fire Prevention Programs

    HF Home tire safety
    RS Residential sprinklers
    SD Smoke detectors

CP   Commercial Property Fire Prevention Programs

PP   Population-specific Fire Prevention Programs

    EC   Elderly Community Resident Fire Prevention and Survival
    LG   Fire Prevention in Languages Other than English

SF   Special-focus Fire Prevention Programs

    AR Arson Prevention
    WL Wildland Fire Prevention
    RF Rural and Farm Fire Prevention
    HC Handicapped

ACTIVE LOCAL NCVFPP PROJECTS

ALABAMA

Organization: **Orange Beach Chamber of Commerce**

Original Contact: Mr. Mike Contorno
Original Address: Orange Beach Chamber of Commerce
    P. O. Drawer 399
    Orange Beach, Alabama 36561

Current Contact: Ms. Kim Bexley
Current Address: Unchanged

Target Audience: Winter visitors to the Gulf Coast (SA, GP)

Problem: The large volume of winter visitors staying in condominiums and hotels on the Gulf Coast during January through April created a need for a fire safety brochure.

Program: A tire-safety brochure was developed and placed in area hotels, condominiums, and at the visitor information center to inform all visitors to the area of the procedures for fire safety and escape in the event of an emergency in commercial lodgings. (CP,PP:EC)

Organization: **University of Alabama Hospital at Birmingham/UAB Burn Center**

Original Contact: Ms. Beverly Bowens
Original Address: University of Alabama Hospital at Birmingham
    UAB Burn Center
    619 S. 19th Street
    Birmingham, Alabama 35233

Current Contact: Unchanged
Current Address: Unchanged

Target Audience: The elderly (SA)

Problem: Elderly citizens have experienced a high risk of burn injury and tire-related death. A significant amount of those burn injuries have been attributable to residential fires, and smoking, cooking, and bathing accidents, as well as accidental ignition of clothing.

Program: The UAB Burn Center undertook development of a tire and burn prevention education program for the elderly covering topics such as cooking safety and fire escape planning. The program includes informational presentations, involving use of films and distribution of fire prevention brochures, at high-rise senior citizen residences and other senior citizen communities. The program also features distribution of fire extinguishers and installation of smoke detectors in homes of the elderly. (PP:EC)

CALIFORNIA

Organization: **Fire Prevention Council**

Original Contact: Mr. Harry Steimer
Original Address: California State Firemen's Association
2701 K Street, Suite 1
Sacramento, California 95823

Current Contact: Ms. Lisa Myers
Current Address: Fire Prevention Council
Fresno Fire Department
450 M Street
Fresno, CA 93721

Target Audience: Children in the first through sixth grades, the elderly, and expectant and young parents (ES, A, SA

Problem: Children in grades one through six have lacked ongoing fire safety education. In addition, many elderly citizens have been especially vulnerable to the threat of fire because of substandard housing.

Program: Volunteers, school officials, and fire department personnel worked together to install smoke detectors in the homes of the elderly. Volunteers conducted seminars for senior citizens on the use of smoke detectors. Burn prevention programs were presented at local hospitals and juvenile medical care facilities. (SB:ES; RB:HF,SD; CP; PP:EC)

Organization:	We Tip, Inc., War on Arson

Original Contact: Original Address:	Ms. Miriam Eckert We Tip, Inc., War on Arson P. 0. Box 740 Ontario, California 91762
Current Contact: Current Address:	Ms. Miriam Brownell We Tip, Inc., War on Arson P. 0. Box 1296 Rancho Cucamonga, California 91729-1296
Target Audience:	General population nationwide, with emphasis on rural areas, which were experiencing an increase in juvenile firesetting (GP, ES, SS)
Problem:	Arson has caused more property losses in California than other major property crimes, including robbery, burglary, and vehicle theft. Both the annual number of arson fires in California and the annual amount of property losses have risen. Fires set for fraudulent insurance claims and tires set by juveniles have been the most common types of arson fires.
Program:	We Tip, Inc., established a 24-hour, seven-days-a-week, toll-free hotline for individuals to report information on deliberately set fires; information gathered through the hotline was used to aid police and fire department investigations of suspicious fires. The organization produced posters and billboards to be placed at arson scenes and neighborhoods with high incidence of arson. Public service announcements, pamphlets, bumper stickers, and other media materials were produced and distributed to increase citizen awareness of arson and the hotline. We Tip, Inc., also presented anti-arson educational programs in primary and secondary schools and at meetings of community organizations. (SB:ES; SF:AR)

FLORIDA

Organization: **Progressive Firefighters Association**

Original Contact: Mr. Roy Jerelds
Original Address: Progressive Firefighters Association
    P. 0. Box 5553
    Orlando, Florida 32855

Current Contact: Unchanged
Current Address: Progressive Firefighters' Association
    P.O. Box 570966
    Orlando, Florida 32857-0966

Target Audience: Elderly, handicapped, low-income, and general adult populations; preschool and elementary school children (PS, ES, SA, GP, SP)

Problem: Most fire deaths and injuries have resulted from residential fires. Specific fire problems have included clothing ignition, careless smoking, and misuse of electrical appliances. Most deaths and injuries have occurred among the elderly and children.

Program: Currently upgrading programs to meet community needs. Teaching fire prevention and survival skills to elementary school, elderly, and handicapped populations. (SB:ES; PP:EC; SF:HC)

LOUISIANA

Organization: **Hammond Fire Prevention Bureau**

Original Contact: Mr. Ronnie Schillace
Original Address: Knights of Columbus
       P. O. Box 2952
       Hammond, Louisiana 70404

Current Contact: Unchanged
Current Address: Hammond Fire Prevention Bureau
       P. O. Box 2788
       Hammond, LA 70404

Target Audience: The general population and elementary school children ages 6 to 12 (ES, GP)

Problem: A lack of education on the importance of smoke detectors and a lack of adequate fire prevention education for young children have left the population vulnerable to fire-related death or injury.

Program: Learn Not to Burn Resource Curriculum. The program featured fire safety classes in schools, hospitals, and libraries. The program also arranged for Boy Scouts to distribute flyers about smoke alarms in local malls and promoted fire safety activities through the news media. (SB:RS; RB:HF)

MAINE

Organization:	**Knights of Columbus**
Original Contact: Original Address:	Mr. Edward A. Minor Knights of Columbus St. John's Council #8144 P.O. Box 1626 Scarboro, Maine 04074
Current Contact: Current Address:	Mr. Gerald S. DiMillo 380 Congress Street Portland, Maine 04101
Target Audience:	Southeast Asians living in inner-city multiple family dwellings (SP)
Problem:	Recent arrivals from Southeast Asia have been vulnerable to fire deaths and injuries due to a lack of education in fire prevention and safety matters.
Program:	Efforts to reach asian immigrants included the conduct of a survey to determine their fire-related needs, development and distribution of fire safety booklets and brochures in Southeast Asian languages, development of a multi-lingual booklet/tape package, production of television public service announcements, and presentations on fire safety. Distribution of materials was conducted through the Refugee Settlement Office and through door-to-door fire inspections. (RB:HF, PP:LG)

MASSACHUSETTS

Organization: **Safety Council of Western Massachusetts**

Original Contact: Mr. Robert L. Petrucelli
Original Address: Safety Council of Western Massachusetts
90 Berkshire Avenue
Springfield, Massachusetts 01109

Current Contact: Mr. James C. Moynihan
Current Address: Unchanged

Target Audience: General adult population (A)

Problem: An increasing number of fires are being caused by carelessness and lack of fire prevention knowledge and fire survival skills. Improper installation and use of wood stoves also contribute to fire deaths and injuries.

Program: The program focused on fire prevention and safety skills education for the general public through video presentations at local organizations, distribution of pamphlets, and public service announcements in newspapers. (RB:HF)

MISSOURI

Organization: **Kirksville Noon Lions Club**

Original Contact: Mr. Stan East
Original Address: Kirksville Noon Lions Club
                      800 West Jefferson
                      Kirksville, Missouri 63501

Current Contact: Unchanged
Current Address: Unchanged

Target Audience: Preschool, elementary school, secondary school children and general population (PS, ES, SS, GP)

Problem: A lack of education on fire prevention and safety. This program begen with an emphasis on PS and ES children and has been expanded to address the needs of older persons.

Program: A mobile "fire safety house" was developed so it could be transported to schools, fairs, etc. It includes a kitchen and second floor bedrooms. Artificial smoke can be produced to provide a realistic experience. An accompanying fire safety and prevention education program also is presented. (SB:ES,PS)

Organization: **The Salvation Army**

Original Contact: Ms. Dianna L. Stenger
Original Address: The Salvation Army
                      101 West Linwood Boulevard
                      P. 0. Box 412577
                      Kansas City, Missouri 64141

Current Contact: Unchanged
Current Address: The Salvation Army
                      3637 Broadway
                      P.O. Box 412577
                      Kansas City, MO 64141

Target Audience: Low income families, inner-city residents and families, elderly, and handicapped (SA, SP)

Problem: Inner-city residents have experienced a high occurrence of fires due to a lack of awareness about fire safety needs in the home, use of alternative heat sources, crowded living quarters, and lack of education.

Program: The program features fire safety presentations to individuals in Salvation Army lodging programs and training of volunteers to give fire safety presentations out in the community and in the home. The program also involves dissemination of fire safety education and installation of smoke detectors to residents through fire safety inspections of low-income homes. (RB:HF,SD; PP:EC)

NEW YORK

Organization: **North East Area Development, Inc. (NEAD)**

Original Contact: Mr. Edward Raskin
Original Address: North East Area Development, Inc.
     1564 East Main Street
     Rochester, New York 14609

Current Contact;   Ms. Nan Aronson
Current Address: Unchanged

Target Audience: The general public and the hearing impaired (GP, SP)

Problem: Individuals living in older structures, senior citizens, children left unattended during the day, recent immigrants, and deaf people have been the segments of the population most susceptible to fire hazards.

Program: The group coordinated a mass public information campaign involving distribution of newsletters and brochures on fire safety issues such as tire escape planning and citizen maintenance of fire hydrants in the winter ("Adopt-A-Hydrant" program); development of a play dramatizing fire safety issues to be aired on cable television; coordination of arson prevention and fire extinguisher workshops; distribution of fire extinguishers and smoke detectors; and identification of homes with defective heating systems. The NEAD also established a neighborhood "Vacant House Watch," including a video assessment of neighborhood vacancies, and developed a specific program for the hearing impaired. (RB:HF,SD; SF:HC)

Organization: **Orleans Community Action Committee, Inc.**

Original Contact: Ms. Mary Ann Wilson
Original Address: Orleans/Genesee Rural Preservation Corporation, Inc.
     409 East State Street
     Albion, New York 14411

Current Contact: Ms. Nancy C. Russell
Current Address: Orleans Community Action Committee, Inc.
     409-411 East State Street
     Albion, New York 14411

Target Audience: Pre-school school children, adults, elderly citizens and special populations (PS, A, SA, SP)

Problem: Rural, low-income, elderly and handicapped citizens residing in old, substandard housing lack adequate fire safety knowledge.

Program: The Housing program provides safety and health-related repairs and services to low-income, elderly and handicapped residents. Burn surveys and outreach programs to spread fire prevention information also are conducted. (SB:ES; RB:HF; SF:RF,HC)

NORTH CAROLINA

Organization: **Carteret Community Action, Inc.**

Original Contact: Mr. Leon Mann, Jr.
Original Address: Carteret Community Action, Inc.
                     Broad and Drawer Streets
                     P. O. Drawer 90
                     Beaufort, North Carolina 28516

Current Contact: Unchanged
Current Address: Carteret Community Action, Inc.
                     216 Turner Street
                     P. O. Drawer 90
                     Beaufort, North Carolina 28516

Target Audience: The elderly and elementary school children (SA, ES), those living alone and those living in mobile homes (ES, SA, SP)

Problem: The majority of fire deaths in the community could have been prevented with better fire prevention education. Most victims, particularly those in mobile homes, did not have smoke detectors in their residences.

Program: The program produced video materials on smoke detector use and maintenance and educational materials for fire prevention presentations by fire departments to children and senior citizens. The program distributed and installed smoke detectors and emergency locator flashing lights in the homes of high-risk individuals. (RB:HF,SD)

Organization: **The Eastern Band of Cherokee Indians**

Original Contact: Ms. Jackie Moore
Original Address: The Eastern Band of Cherokee Indians-Community Injury Control
                     P. O. Box 666
                     Cherokee, North Carolina 28719

Current Contact: Ms. Charlene Toineeta
Current Address: Unchanged

Target Audience: Residents of the Cherokee Indian reservation in North Carolina (GP)

Problem: Lack of smoke detectors and home fire escape plans and improper use of wood stoves have contributed to tire losses and injuries.

Program: The group developed an educational program for elementary school children and community groups. The program specifically addressed use of fire escape ladders and tire extinguishers. In addition, a fire safety video was broadcast on cable television. (SBES; RB:HF)

Organization: **Onslow County Cooperative Extension Service**

Original Contact: Mr. J. Gregory Clemmons
Original Address: Onslow County Agricultural Extension
                           Onslow County 4-H Club
                           604 College Street, Room 8
                           Jacksonville, North Carolina 28540

Current Contact: Unchanged
Current Address: Onslow County Cooperative Extension Service
                           604 College Street, Room 8
                           Jacksonville, NC 28540

Target Audience: Children up to 14 years old (ES)

Problem: Increases in the number of latch-key children, use of alternative heating devices and cooking devices have contributed to a high incidence of fires in residential settings.

Program: During the month of October, youths between the ages of nine and 14 are instructed in fire prevention techniques in a workshop format. Volunteer fire department members are involved in the instructional process using their resources and the 4-H curriculum available on fire safety. At the conclusion of the sessions, a 4-H Fire Quiz Bowl is held. (SBES)

NORTH DAKOTA

Organization:	**Roughrider Country Kiwanis**

Original Contact: Mr. Calvin Lundberg
Original Address: Roughrider Country Kiwanis
646 9th Avenue West
Dickinson, North Dakota 58601

Current Contact: Mr. Terry Wehner
Current Address: 633 Ninth Ave. West
Dickinson, North Dakota 58601

Target Audience: The general population, especially residents using alternative heating sources such as wood-burning stoves, liquid fuel heaters, and electric heaters (GP)

Problem: Fire losses and injuries have increased as a result of improper installation and misuse of alternative heaters, poor maintenance of chimneys, proximity of solid fuels to heating devices, lack of awareness of tire hazards, and lack of residential sprinkler systems.

Program: Volunteers and staff built and displayed a mobile demonstration unit for fire safety education. The unit was used to show audiovisual education programs; proper installation, use, and maintenance of alternative heating devices; use and maintenance of smoke detectors; and printed literature on fire safety in the home. The program also featured a public awareness media campaign and presentations throughout the region to educate the public on behaviors intended to prevent residential fires. (RB:HF,SD)

OKLAHOMA

organization: **American Red Cross, Tulsa Area Chapter**

Original Contact: Ms. Carol Lofton
Original Address: American Red Cross, Tulsa Area Chapter
10151 E. 11th Street
Tulsa, Oklahoma 74128

Current Contact: Mr. Bob E. Roberts
Current Address: Unchanged

Target Audience: Residents of the 16 Tulsa Housing Authority (THA) complexes, area elementary schools, and students served by Indian Community Health Services in the area (SP, ES)

Problem: A high incidence of fires in low income housing has resulted in needless deaths and injuries. The Tulsa Fire Department has attributed these fires to a general lack of awareness of fire safety, with the most common fire causes recorded as smoking accidents, cooking or kitchen fires, juvenile fire play, and heating equipment.

Program: A preventative education program focused on protective measures, preventative behavior, and survival techniques. The Tulsa Red Cross developed and distributed printed fire prevention safety and education materials to complex residents; held fire safety demonstrations and seminars for tenants; conducted a mass media campaign, including video tapes, to promote adult and child awareness of fire hazards; conducted media demonstrations of survival techniques; developed fire safety guides for elderly or handicapped individuals and for babysitters; and trained Creek and Cherokee Indian Community Health Representatives as outreach instructors (SB:ES; RB:HF, PP:EC; SF:HC)

PENNSYLVANIA

Organization: **Allegheny Township Fire Prevention, Inc.**

Original Contact: Mr. Bruce Kaufman
Original Address: R. D. #2, Box 373
     Altoona, Pennsylvania 16601

Current Contact: Mr. David J. Blair
Current Address: Unchanged

Target Audience: The general population and mentally retarded groups (GP, SP)

Problem: A lack of smoke detectors in homes and improper use of wood stoves and supplemental heating supplies have contributed to area fire problems. Also, educating the general public on reporting emergency situations to a dispatcher is necessary.

Program: The program developed a mobile fire safety education unit consisting of a three-room trailer, with one room equipped with displays on smoke detector installation and maintenance; a second room equipped with displays on wood stove installation and maintenance;, and a third room used for training participants in home fire escape and emergency reporting. The program also includes a brief film on reacting to fire in the home. The mobile program has been presented at area schools, youth group meetings, fire company open houses, community group meetings, and meetings of mentally retarded citizen groups. (RB:HF; SF:HC)

organization: **Associated Services for the Blind**

Original Contact: Mr. Vincent M. McVeigh
Original Address: Associated Services for the Blind
     919 Walnut Street
     Philadelphia, PA 19107

Current Contact: Unchanged
Current Address: Unchanged

Target Audience: Special Populations (SP)

Problem: The visually impaired are particularly vulnerable to fire death, burn injury, and smoke inhalation. Visual impairments may hinder response time and place people at greater risk in tire emergency situations.

Program: The Associated Services for the Blind produced a video cassette that focused on fire safety for visually impaired persons (VIPs) and training people to work with VIPs. The program included Fire Smart!! workshops in modular form, oral presentations, and brochures outlining workshop highlights. Workshop oral presentations were distributed on audio cassette in suitable format for radio public service announcements. (SF:HC)

Organization:	**The Burn Prevention Foundation**

Original Contact:	Ms. Sandra C. Raymond
Original Address:	Burn Foundation
	East/Northeast Pennsylvania
	1251 South Cedar Crest Boulevard
	Allentown, Pennsylvania 18103

Current Contact:	Mr. B. Daniel Dillard
Current Address:	Burn Prevention Foundation
	5000 Tilghman Street, Suite 110
	Allentown, PA 18104

Target Audience: Senior adults (SA)

Problem: Older adults account for one-third of all fire and burn deaths in the U. S. Despite these national statistics, there has been little focus on fire and burn prevention for this population.

Program: The Burn Prevention Foundation has developed and tested a fire and burn safety educational program that combines elements of prevention awareness, safety, and survival information. Four program units address home hazards, cooking and kitchen safety; smoke detectors; and fire survival at home and in hotels. Volunteers and professional educators have found the program materials easy to use and well received by older audiences (PP:EC)

Organization:	**The Burn Prevention Foundation**

Original Contact:	Mr. B. Daniel Dillard
Original Address:	Burn Prevention Foundation
	5000 Tilghman Street, Suite 110
	Allentown, PA 18104

Current Contact:	Unchanged
Current Address:	Unchanged

Target Audience: Special Populations (SP)

Problem: The nation's 34 million citizens with disabilities are particularly vulnerable to fire death, burn injury, and smoke inhalation. Mobility impairments may hinder response time and place people at greater risk in fire emergency situations.

Program: The Burn Prevention Foundation has developed an educational package that addresses the special fire and burn risks for those with mild to moderate mobility impairments. Included in each program kit are: a 20-minute video and reproducible masters for supplemental brochures; a self-test; and a discussion leaders guide. Fire safety and survival issues addressed include escape planning and when escape is not feasible, sheltering, at home and in the workplace. Particular emphasis has been placed on hazards associated with smoking, cooking from a wheelchair, and the prevention of tap water scalds. Protective initiatives stressed include the value of preplanned fire escape routes or sheltering, modification of clothing for safer cooking, alternate cooking and food serving methods, safe smoking practices, and burn first aid. (RB:HF; SF:HC)

Organization:	**Museum of Scientific Discovery**
Original Contact:	Mr. Alan H. Ruppert
Original Address:	Museum of Scientific Discovery
	P. 0. Box 934
	Strawberry Square
	Harrisburg, Pennsylvania 17108
Current Contact:	Mr. Steven C. Ling
Current Address:	Unchanged
Target Audience:	Elementary school children (Es)
Problem:	Many victims of residential fires in the area are elementary school children.
Program:	The program developed an interactive computer software exhibit that creates an individualized plan of each child's bedroom. Group presentations are followed up with fire safety assessments in the children's homes. (SB:ES)

SOUTH DAKOTA

organization:	**Keep South Dakota Green Association (KSDGA)**
Original Contact:	Mr. David W. Erickson
Original Address:	Keep South Dakota Green Association
	P. 0. Box 3
	Pierre, South Dakota 57501
Current Contact:	Unchanged
Current Address:	Keep South Dakota Green Association
	P. 0. Box 3
	Pierre, South Dakota 57501-0003

Target Audience: Residents of the rural Black Hills area, particularly those in areas where timberland is being developed for housing (GP)

Problem: Wildfires are occurring within close proximity of rural houses located in forested areas. "Fuel loading," or the accumulation over time of logging debris, pine needles, dead grasses, and other fuels, in forests and on and around homes, in addition to increased numbers of visitors to wilderness areas, were identified as potential fire hazards.

Program: Facilitators informed residents of hazards and basic fire safety procedures around rural homes; developed and distributed brochures and public service announcements targeting residents of subdivisions bordering wildland areas; encouraged residents to practice fire safety techniques through public service announcements, symposiums, and brochures; developed a video of an actual forest fire; and acquired and used Smokey the Bear costumes for parades and classroom visits. Volunteers also conducted conferences on the local fire problem that included local and state policymakers. A supply of the publication "Wildfire Safety Guidelines for South Dakota Rural Homeowners," prepared by KSDGA and the Division of Forestry is available. Copies of the publication are provided to interested homeowners upon request. (RB:HF; SF:WL,RF)

Organization:	**Marion Area Jaycees/Marion Area Senior Citizens' Center, inc.**

Original Contact: Mr. Dick Luke
Original Address: Marion Area Jaycees
                         Box 226
                         Marion, South Dakota 57043

Current Contact: Unchanged
Current Address: Unchanged

Target Audience: Senior citizens, elementary school children, parents of young children, and members of civic groups (ES, A, SA)

Problem: Fire injuries and losses can be attributed largely to lack of knowledge and awareness of fire safety issues, lack of escape plans for homes and facilities, and lack of smoke detectors in some homes.

Program: Firefighters conducted educational sessions for first through sixth graders in which students learned about fire prevention and survival and received training in helping their families to design home escape plans. With the help of volunteers, fifth graders produced fire safety activity books and videotape presentations on home and farm fire safety. Programs on fire and burn prevention also were presented to adults through parents' nights at schools and through fire department open houses and civic group meetings. Senior citizens also received fire and burn safety education. In addition, the program distributed smoke detectors throughout the community and educated recipients on use and maintenance of the devices. (SB:ES; RB:HF,SD; PP:EC)

Organization:	**Retired Senior Volunteer Program of Beadle County**

Original Contact: Ms. Ann Sibley
Original Address: Retired Senior Volunteer Program of Beadle County
                         7th and Ohio S.W.
                         Huron, South Dakota 57350

Current Contact: Ms. Darlene French
Current Address: Unchanged

Target Audience: Elderly residents and elementary school children (ES, SA)

Problem: Fires resulting from heating units, cooking, arson, and faulty electrical wiring as well as a lack of fire safety awareness, lack of smoke detectors, and lack of data on burns, especially among the elderly population, have been identified as problems. Cooperation of schools also has been a problem.

Program: Retired senior volunteers and volunteer firemen distributed locally developed fire safety information in seminars targeting civic groups, elementary school students, and elderly residents; purchased and installed smoke detectors and new batteries; through the use of media and public service campaigns educated children and the public on fire escape routes; and sponsored a fire safety poster contest. (SB:ES; RB:HF,SD; PP:EC)

TENNESSEE

Organization:	**Town of Collierville Fire Department**

Original Contact:
Original Address:

Mr. Ben F. Wilson
Chamber of Commerce
The Depot
Town Square
Collierville, Tennessee 38071

Current Contact:
Current Address:

Unchanged
Mr. Ben F. Wilson
Town of Collierville Fire Department
P. O. Box 636
Collierville, TN 38027-0636

Target Audience: Preschool, Elementary school children and the general public (PS, ES, GP)

Problem: Because many new residents have little fire prevention knowledge, rapidly growing residential areas have a high risk of fires resulting in death or serious injury.

Program: Fire safety education for children included an in-school fire prevention curriculum, home exit drills, and presentations about juvenile firesetting and about firefighting careers. Babysitters received instruction in first aid, fire prevention, and fire survival. General public awareness programs include seasonal tire prevention activities, fire safety booths at local fairs, distribution of monthly newsletters and other fire safety literature, radio broadcasts of public service announcements, and demonstrations of residential sprinkler systems at local home shows. In addition, smoke detectors and fire extinguishers are installed and maintained in the homes of elderly, handicapped, and low-income citizens. (SB:ES; RB:HF,RS,SD)

TEXAS

Organization:	**American Red Cross, Beaumont Chapter**

Original Contact:
Original Address:
Ms. Teresa Johnson
American Red Cross, Beaumont Chapter
706 Magnolia
Beaumont, Texas 77701

Current Contact:
Current Address:
Mr. Tom Martin
Unchanged

Target Audience: Residents in low-income minority areas at high risk of fire, schoolchildren, and the elderly (ES, SA, SP)

Problem: Many residential fires are caused by heating, cooking, and smoking accidents. Only a small percentage of minority area residences have smoke detectors.

Program: "Project Save a Home" developed informational materials on fire prevention, risk reduction, and installation of smoke detectors for home owners and renters and distributed these educational materials to high risk residents through various locations in the community: the local food stamp office; Meals on Wheels; and senior citizen centers for the elderly. Volunteers make presentations to senior citizens at area senior centers, Beaumont Nutrition Centers, local schools, and libraries. (RB:HF,SD;CP)

Organization:	**Post Chamber of Commerce**

Original Contact:
Original Address:
Mr. Lewis H. Earl
Post Economic Development Corporation
P. O. Box 610
Post, Texas 79356

Current Contact:
Current Address:
Ms. Dru Ann Laus
Post Chamber of Commerce
P. O. Box 610
Post, Texas 79356-0610

Target Audience: Rural residents and senior citizens (SA, GP)

Problem: Most area residents have not had training in fire prevention, fire and burn safety, storage of combustibles, exit planning, or emergency behavior. Many homes and workplaces do not have smoke detectors or fire extinguishers. The low rainfall in the area has contributed to fire damage. In addition, fire department response times are longer than average because the area is rural.

Program: The program involved development and presentation of informational lectures on rural fire hazards, home safety, and evacuation to a variety of community and civic organizations. The programs included fire prevention films and handouts on fire prevention and burn safety. Volunteers also distributed and installed smoke detectors in the homes of the elderly and the economically disadvantaged. (RB:HF,SD; PP:EC)

UTAH

Organization:	**Your Community Connection (YCC)**
Original Contact:	Ms. Gaye Littleton
Original Address:	Ogden YWCA
	505 27th Street
	Ogden, Utah 84119
Current Contact:	Unchanged
Current Address:	Ms. Gaye Littleton
	Executive Director
	Your Community Connection (YCC)
	2261 Adams Avenue
	Ogden, UT 84401

Target audience: Female-headed, single-parent families and low-income residents (PS,ES,A)

Problem: In northern parts of the state, including inner-city areas, elementary school children often remain unattended at home while their mothers work. A lack of smoke detectors and fire extinguishers and general unawareness of tire safety issues also increase fire risks.

Program: The YWCA designed a fire prevention and awareness program for female single parents and their children. In addition, the YWCA provided for the distribution of smoke detectors to single-parent families and low-income residents and the education of smoke detector recipients in maintenance of the detectors. Educational information was transmitted through advertisements, pamphlets and brochures, and seminars specifically geared toward selected audiences.   (RB:HF,SD)

VIRGINIA

Organization:	**Loudoun Business and Professional Women's Club**

Original Contact:
Original Address:

Ms. Laura Pearson
Loudoun Business and Professional Women
P. 0. Box 157
Philomont, Virginia 22131

Current Contact:
Current Address:

Mr. J. B. Anderson
16600 Courage Ct.
Leesburg, Virginia 22075

Target Audience:

Preschool, Elementary, Secondary School, Adult, Senior Adult, Mentally and Physically Challenged (PS, ES, SS, A, SA, GP, SP)

Problem:

Many residents movong into suburban or rural areas lack knowledge of the slower response times of firefighters and E.M.S. units to accident scenes in these areas. Uncontrollable fires result in higher dollar loss, injuries, and death.

Program:

We have expanded our program delivery to include programs at all levels. The program targets schools, businesses, senior citizen sites, and adult groups. Program deliverables include brochures and handouts on life safety and built-in protection systems; a resident emergency services awareness package mailed to current and new residents; safety and fire extinguisher training for businesses; and special programs, such as McGruff Safety Camp and the Sesame Street Safety Show. (SB:ES,PS; RB:HF,RS,SD; CP; PP:EC; SF:AR,RF,HC)

WEST VIRGINIA

Organization: **Charleston Chamber of Commerce**

Original Contact: Mr. John Chapman
Original Address: Charleston Chamber of Commerce
                      818 Virginia Street East
                      Charleston, West Virginia 25301

Current Contact: Unchanged
Current Address: Charleston Regional Chamber of Commerce
                      106 Capitol Street
                      Suite 100
                      Charleston, WV 25301-2610

Target Audience: The general population, merchants, commercial property owners, the elderly, and children (Es, SA, GP)

Problem: Fire deaths among the elderly, especially the infirm and the handicapped, and among the very young, seasonal tires; careless leaf burning, unsafe use of alternative heaters; and vacant downtown buildings are local fire problems.

Program: Program administrators and volunteers developed and presented a training and awareness program featuring video films, media programs, and seminar presentations, which included discussion on preparing and using an evacuation plan, preparing for the arrival of the fire department, using fire extinguishers, training employees for fire safety, and encouraging fire safety in commercial structures. The program is now certifying users of buildings comprising at least 20,000 square feet rather than 50,000 square feet as had been required at the beginning of the program. The Life Safe Committee now certifies buildings in these cities: St. Albans, Dunbar, South Charleston, and Charleston. As of 1992, a total of 91 buildings had been certified. (RB:HF; CP)

APPENDIX 1

ADDITIONAL LOCAL NCVFPP PROJECTS

ALABAMA

Organization: **Civitans**

Original Contact: Mr. Allen Hawkins
Original Address: Civitans
                90 Broad Street
                Gadsden, Alabama 35999

Current Contact: Mr. Dan Wescott
Current Address: Civitans
                322 Walnut St.
                Gadsden, Alabama 35903

Target Audience: Youths, especially those under nine years of age, and their parents (ES, A)

Problem: Although fire-related deaths and injuries have declined overall, fire-related deaths and injuries among children have risen. Most of those deaths and injuries have resulted from household fires caused by improper use of appliances, electrical equipment, smoking materials, and flammable materials. In addition, many homes have not been equipped with smoke detectors.

Program: The Civitans arranged educational fire safety presentations in elementary schools, distributed smoke detectors and informational literature on fire safety, and sponsored development of broadcast public service announcements to increase public awareness of fire safety issues, especially those concerning children. (SB:ES; RB:HF,SD)

Organization: **Exchange Club of Huntsville**

Original Contact: Mr. Mike Edwards
Original Address: The Exchange Club of Huntsville
                1019 Putman Drive
                Huntsville, Alabama 35816

Current Contact: Unchanged
Current Address: Unchanged

Target Audience: The elderly and handicapped populations (SA, SP)

Problem: A significant proportion of fire deaths and injuries have occurred among the elderly and handicapped populations. Those two population groups have faced an increased risk of fire-related injury because they have lacked the knowledge and skills needed to respond quickly in fire emergencies and to make effective use of fire protection equipment.

Program: Public service announcements were produced to increase community awareness of fire safety problems of the elderly and to make the elderly aware of the availability of fire safety training and equipment. Brief videotape and slide presentations on fire safety measures and the use of fire protection equipment were developed, produced, and presented to elderly and handicapped populations. Smoke detectors and fire extinguishers were distributed to elderly and low income citizens. (PP:EC, SF:HC)

ALASKA

Organization:	**Anchorage Firefighters Union**

Original Contact:	Mr. Ron Ozmina
Original Address:	Anchorage Firefighters Union
	1200 East 76th Ave.
	Suite 1227
	Anchorage, Alaska 99618

Current Contact:	Mr. Thomas Preston
Current Address:	Unchanged

Target Audience: Elementary school children and the general public (ES, GP)

Problem: Anchorage schools lacked the organization and expertise necessary to effectively present a fire safety curriculum. In addition, the city lacked sufficient fire data to define specific fire problems in the area.

Program: The firefighters' union helped schools establish a fire safety curriculum program. A tracking system was established to gather statistics about local fires and identify specific local fire problems. After specific fire problems were determined, public service announcements were produced to increase public awareness of those problems. In addition, a local grocery store aided the fire prevention program by distributing smoke detectors to the elderly, disabled, and families with infants. The local fire department provided installation assistance and home fire safety inspections to detector recipients. (SB:ES; RB:HF)

Organization:	**B. P. 0. Elks No. 1429**

Original Contact:	Mr. Steve Rydeen
Original Address:	B. P. 0. Elks No. 1429
	P. 0. Box 5177
	Ketchikan, Alaska 99901

Current Contact:	Mr. Walt Smith
Current Address:	Unchanged

Target Audience: General suburban population, with focus on children (ES, GP)

Problem: Although Alaska has one of the smallest state populations, fire fatalities and property losses have exceeded the national average. The state's harsh environment, the low quality of many homes and cabins, and a lack of fire education have contributed to the problem. Common causes of the fires have included faulty heating systems and improper use of smoking materials.

Program: The Ketchikan Elks Lodge purchased videotapes, pamphlets, and other educational materials. With the assistance of local fire departments, the Elks trained local public school teachers in fire prevention and safety to enable those teachers to present the curriculum materials to students in seven community school systems. Public service announcements and printed advertisements also were produced to increase fire safety awareness among the general population. (SB:ES; RB:HP)

Organization:	**Yukon-Kuskokwim Health Corporation**
Original Contact:	Ms. Mary Wilda Warner
Original Address:	Yukon Kuskokwim Health Corporation
	P. O. Box 528
	Bethel, Alaska 99559
Current Contact:	Unchanged
Current Address:	Unchanged
Target Audience:	School children and parents (ES, A)
Problem:	Fires have been one of the leading causes of accidental death in the Yukon-Kuskokwim Delta area. Faulty heating systems and misuse of smoking materials have been among common causes of accidental fires.
Program:	Public service announcements in English and in Yup'ik, the local native language, were aired regularly on the local television and radio stations. A videotape presentation outlining common causes and prevention measures for fires in village homes was produced in both languages. The training videotape was presented in schools, on television, and through public meetings of adult citizens. A fire education program was established in public schools. The community held a "Smoke Detector Week," which included such activities as a local fire department open house, a poster contest for schoolchildren, a series of fire extinguisher demonstrations, and the sale of smoke detectors. (SB:ES; RB:HF; PP:LG)

CALIFORNIA

Organization:	**Covina Women's Club**

Original Contact:	Ms. Jan Gratton
Original Address:	Covina Women's Club
	Safety Department
	128 South San Jose Street
	Covina, California 91723

Current Contact:	Unchanged
Current Address:	Covina Women's Club
	Safety Department
	469 South Albertson Avenue
	Covina, California 91723

Target Audience: Junior and senior high school students (SS)

Problem: Adolescents have been exposed to an increased risk of burn from several activities, including working near hot liquids or grease in restaurants, suntanning, working on automobiles and motorcycles, and careless cigarette smoking. However, adolescents often have lacked adequate fire safety knowledge because most recently instituted fire education programs have concentrated on the elderly or young children.

Program: The Women's Club, with the assistance of the local fire department, developed and administered a survey to evaluate the fire knowledge and attitudes of high school students and determine what type of fire education would be most appropriate for the students. The information obtained from the survey was used to make recommendations for the design of a fire and burn prevention curriculum. (SB:ES)

DELAWARE

Organization: **City Side, Inc./Deiaware First**

Original Contact: Mr. Sam Beard
Original Address: Delaware First
                       618 Market Street Mall
                       Wilmington, Delaware 19801

Current Contact: Ms. Michele Rossi
Current Address: City Side, Inc.
                       c/o Delaware First
                       206 West Tenth Street
                       Wilmington, Delaware 19801

Target Audience: Children, the elderly, new home owners, non-English speaking Hispanic residents, and the general population (PS, ES, SP, GP)

Problem: Fires have been a leading cause of death among children in Delaware; furthermore, the rising number of working parents has left many children in the care of day care providers lacking adequate fire safety knowledge. Lack of communication between fire departments and the public on such seasonal fire safety issues as holiday decorating and chimney cleaning has presented an additional fire problem.

Program: In order to introduce a fire safety curriculum in day care centers and elementary and middle schools, the program provided training and curriculum materials to teachers and day care providers. The program provided funding assistance to local youth-serving organizations for locally designed fire safety projects targeting youth groups. Participatory activities for children included art contests, construction of "fire safe houses," and fire survival drills. General community outreach programs involved workshops held in cooperation with neighborhood groups, as well as distribution of fire safety literature through direct mail and mass distribution at community events. Television and radio public service announcements informed local residents about holiday season fire hazards and the dangers of kerosene heating and careless smoking. Special outreach programs targeted the elderly, and Spanish-speaking populations. (SB:PS,ES; RB:HF; PP:EC,LG)

Organization: **Frederick Douglass Intermediate School Parent-Teacher Association**

Original Contact: Ms. Ellouise H. Martin
Original Address: Frederick Douglass PTA
                         408 Market Street
                         Seaford, Delaware 19973

Current Contact: Ms. Connie Chapis
Current Address: Frederick Douglass PTA
                         Route 1, Box 209C
                         Seaford, Delaware 19973

Target Audience: Rural-area residents, children, and the elderly (GP, ES, SA)

Problem: Because Seaford is a large rural area, fire department response times often have been lengthy. Thus, citizens need to know basic fire prevention and safety techniques in order to avoid and survive fires. In particular, many fires have been caused by the misuse of such alternative heating devices as wood stoves and kerosene heaters and by careless smoking.

Program: The PTA distributed fire safety curriculum materials in elementary schools and arranged fire safety workshops for teachers and adults. The program utilized television, radio, and newspaper advertisements to spread information about local fire problems and developed a cartoon character to highlight the dangers of careless smoking. The program also distributed smoke detectors to low-income, elderly, and handicapped individuals. In cooperation with the local fire department, the program established a mobile "Children's Fire Safety House" to teach youngsters and their parents about fire escape planning. (SB:ES; RB:HF,SD; PP:EC; SF:RF)

Organization: **Milford Jaycees, Inc.**

Original Contact: Mr. Gary Emory
Original Address: Milford Jaycees, Inc.
                         P. O. Box 9
                         Milford, Delaware 19963

Current Contact: Unchanged
Current Address: Milford Jaycees, Inc.
                         4 South Washington Street
                         Milford, Delaware 19963

Target Audience: Schoolchildren and the general population (ES, GP)

Problem: Use of alternative heating devices, cooking fires, clothing ignition accidents, and smoke detectors were identified as issues of special concern to area residents.

Program: The Jaycees worked to establish fire safety curricula in schools and open fire safety seminars throughout the community. Materials highlighting fire safety issues were distributed at community fairs and festivals. Public service announcements were produced and broadcast on radio and television. Smoke detectors were distributed to low-income residents. A mobile "Fire Safety House" was built for use in demonstrations of fire safety techniques throughout the community. (SB:ES; RB:HF,SD)

DISTRICT OF COLUMBIA

Organization:	**College Achievement Module, Inc.**

Original Contact: Dr. John T. Blue
Original Address: College Achievement Module, Inc.
University of the District of Columbia
4200 Connecticut Avenue, N. W.
Washington, D. C. 20008

Current Contact: Unchanged
Current Address: 904 Northwest Drive
Silver Spring, Maryland 20901

Target Audience: Residents of the District of Columbia's ethnically diverse Ward One (SP)

Problem: Fires have contributed to the decay of inner city neighborhoods by depleting the stock of low- and moderate-income housing, prompting business owners to relocate, and destroying buildings of historic significance. A lack of smoke detectors and adequate fire safety knowledge among the general public have contributed to the fire problem.

Program: The program enlisted youth involved in summer youth work programs in the District to survey Ward One residents about their compliance with fire safety codes, use of smoke detectors in their homes, and knowledge of home fire safety measures. The youth provided survey respondents with information about home fire safety measures and encouraged respondents to practice fire safety in the home and teach other family members about fire safety. Because of the large Hispanic population in the ward, the program employed youths who could speak Spanish and included Spanish-language literature. (RB:HF,SD; PP:LG)

FLORIDA

organization: **Port Orange-South Daytona Chamber of Commerce**

Original Contact: Ms. Gigi Boileau
Original Address: Port Orange-South Daytona Chamber of Commerce
                     3431 Ridgewood Avenue
                     Port Orange, Florida 32019

Current Contact: Unchanged
Current Address: Unchanged

Target Audience: Residential populations of Port Orange and South Daytona (GP)

Problem: In Port Orange and South Daytona, the percentage of all structural fires occurring in residences has been higher than state and national averages. The most common causes of residential fires have been cooking accidents, careless smoking, and misuse of electrical equipment.

Program: Program activities included informational presentations, fire safety training, and distribution of informational materials and fire safety equipment. Activities took place at community events, including fairs and community group meetings. Specific activities included video and slide presentations at a county fair, public demonstrations by fire department personnel of fire extinguisher use, distribution of smoke detectors, and home fire safety inspections. Public service announcements for newspapers, radio, and television were produced. (RB:HF,SD)

Organization: **Urban League of Greater Miami, Inc.**

Original Contact: Ms. Jacquelyn Rowe-Dottin
Original Address: Urban League of Greater Miami, Inc.
                     8500 N. W. 25th Avenue
                     Miami, Florida 33147

Current Contact: Mr. T. Willard Fair
Current Address: Unchanged

Target Audience: Preschool, elementary school, and secondary school students (PS, ES, SS)

Problem: Youths five to 15 years old have been at substantial risk for burn injuries. Burns suffered by younger children have been predominantly scald burns, while burns suffered by older youth have tended to be flame and grease burns. Risk of burn injury has been highest among minority and low-income youth.

Program: The Urban League worked to deliver the fire prevention message to youth in a variety of ways through the cooperation of fire authorities, the business community, and the media. Firefighters visited schools to teach students of all ages about fire prevention and survival. A major local department store hosted a party for youths that highlighted fire safety. With the cooperation of local television and radio stations, the league produced television specials; public service announcements in English, Spanish, and Creole; and a rap music song and video written by a local student to increase youth awareness of fire issues. With the assistance of local high school students, a fire safety coloring book was created. (SB:ES,PS; PP:LG)

GEORGIA

Organization: **The Exchange Club of Columbus**

Original Contact: Mr. Don Baldwin
Original Address: The Exchange Club of Columbus, Inc., Service Club
                         P. 0. Box 1466
                         Columbus, Georgia 31902

Current Contact: Unchanged
Current Address: Unchanged

Target Audience: Elderly owners of single family homes and elementary age school children (ES, SA)

Problem: A high percentage of tires has resulted from defective wiring, stoves, and furnaces in single-family dwellings occupied by elderly persons. Also, there has been a high incidence of fires from children playing with matches or lighters around combustible materials.

Program: The program featured installation of more than 3,000 smoke detectors, primarily in homes of the elderly. Also, presentations were made to elementary school children using "Pluggie," a talking robotic fire hydrant. (SB:ES; PP:EC, RB:SD)

Organization: **Junior Service League of Rome, Inc.**

Original Contact: Ms. Janet Griffin
Original Address: Junior Service League of Rome, Inc.
                         P.O. Box 1003
                         Rome, Georgia 30161

Current Contact: Ms. Judy Watters
Current Address: Unchanged

Target Audience: Low-income individuals, the elderly, preschool and elementary school children, the handicapped, and general population (PS, ES, SA, SP, GP)

Problem: A lack of smoke detectors in low-income areas contributed to unusually high death rates for elderly, handicapped, and low-income groups.

Program: The program featured distribution of smoke detectors and batteries, installation of a sprinkler system in a home for the mentally ill, and presentations on fire safety for community organizations. The program also included installation of residential sprinkler systems for public review and fire prevention and survival presentations at preschools and elementary schools. Presentations involved use of puppets, movies, videos, and brochures. (RB:HF,RS,SD; SB:PS,ES)

Organization:	**Professional Secretaries International**
Original Contact:	Ms. Lou Ellen Brumbeloe
Original Address:	Professional Secretaries International
	P. O. Box 71
	West Point, Georgia 31833
Current Contact:	Unchanged
Current Address:	Unchanged
Target Audience:	Elementary and secondary school youth, low-income elderly, and the general population (ES, SS, SA, GP)
Problem:	A lack of fire safety knowledge among the elderly, low-income families, and families with small children has increased the risk of fire injury or death and property damage.
Program:	The program featured fire safety presentations in local schools and at meetings of civic and church groups. Materials created by the program included a 12-minute videotape on home fire safety, a new workbook for a tire safety curriculum program for the entire school system, and public service announcements. (SB:ES; RB:HF)

IDAHO

Organization:	**Greater Kootenai County Fire Prevention Cooperative**
Original Contact:	Mr. Bill Severson
Original Address:	Greater Kootenai County Fire Prevention Cooperative
	1712 Golf Course Road
	Coeur d'Alene, Idaho 83814
Current Contact:	Mr. John Hurley
Current Address:	Unchanged
Target Audience:	Elementary school students, elderly and handicapped citizens, and wood stove owners (ES, SA, GP, SP)
Problem:	Most local fires have occurred in buildings without smoke detectors. Elderly and handicapped citizens have been especially vulnerable to death or injury from fires because of their reduced ability to respond quickly to fire emergencies, children have not possessed adequate knowledge of home fire prevention and survival measures, and more than one-fourth of local residential fires have been caused by sparks, flames, or heat from wood stoves.
Program:	The cooperative arranged distribution of smoke detectors and brochures on home fire safety to elderly and handicapped persons and offered assistance with installation and maintenance of smoke detectors. The program also featured fire safety presentations at meetings of senior citizen organizations. The cooperative supported design and implementation of a fire prevention curriculum based on local fire problems, especially wildland fire prevention, in elementary schools. For wood stove owners, the cooperative coordinated distribution of brochures on proper installation, use, and care of wood stoves at home and garden shows, county fairs, civic group meetings, fire departments, and state and federal offices and arranged for the loan of chimney-sweeping kits to wood stove owners to prevent dangerous creosote buildup. In addition, the program used public service announcements and National Fire Prevention Week to increase public awareness of the various tire safety programs. (SB:ES; RB:HF,SD; PP:EC, SF:WL,HC)

Organization:	**Kiwanis Club of Capital City**

Original Contact: Original Address:	Mr. Robert Aldridge Kiwanis Club of Capital City P. O. Box 82 Boise, Idaho 83701
Current Contact: Current Address:	Mr. Leon Grisham Unchanged
Target Audience:	Elementary school students in first, third, and fifth grades (ES)
Problem:	Although juveniles either have caused or suffered injury in most of the fires in the county, no fire education program has existed in area schools.
Program:	The Kiwanis Club, in cooperation with local educators, designed a fire prevention and survival curriculum that highlighted fire safety issues of local import. The curriculum was implemented in the first, third, and fifth grades in county elementary schools. One school held a fire prevention poster contest. The club also sponsored the placement of smoke detectors into elementary school classrooms and the production of television public service announcements to increase public awareness of fire safety issues. (SB:ES)

Organization:	**Malad Lions Club**

Original Contact: Original Address:	Mr. Jack Allred Malad Lions Club Malad, Idaho 83252
Current Contact: Current Address:	Unchanged Unchanged
Target Audience:	The elderly, especially those living alone; school children in the first, third, and fifth grades; and homeowners (SA, ES, GP)
Problem:	Senior citizens, especially those living alone, have been at high risk for fire death or injury. In addition, juveniles have caused or have been injured by numerous fires each year in the county. However, no fire safety education programs have existed in the county.
Program:	The Lions Club program involved development of a series of five presentations to instruct senior citizens about clothing ignition; installation, use, and care of smoke detectors; home fire hazards, including cooking fires; and fire survival. The presentations were given at meetings of senior citizen groups. The club sponsored installation of smoke detectors in homes of the elderly and needy and provided fresh batteries a year after installation. Local educators assisted with the integration of fire prevention and safety education into existing curricula in the first, third, and fifth grades in county schools. The club developed a program to loan chimney-sweeping equipment to local homeowners. Newspaper pubic service announcements were produced to increase public awareness of the availability of the fire safety programs. (SB:ES; PP:EC, RB:HF)

ILLINOIS

Organization:	**American Red Cross, Central Illinois Chapter**
Original Contact: Original Address:	Mr. Marvin Miles American Red Cross, Central Illinois Chapter 1224 N. Berkeley Peoria, Illinois 61603
Current Contact: Current Address:	Unchanged Unchanged
Target Audience:	The general population, with special focus on children, senior citizens, and low-income families (GP, ES, SS, SA)
Problem:	Five major causes of local fires have been identified: children playing with smoking materials; misuse of smoking materials by the elderly; faulty wiring in many homes; multiple use of extension cords, and use of such alternative heating equipment as wood stoves and kerosene heaters.
Program:	The Red Cross arranged informational talks and demonstrations by fire department personnel and other volunteers for all age groups in various settings, including schools, churches, hospitals, formal community events, senior citizen group meetings, and fire department open houses. Presentations for children employed a puppet show, movies with fire-related themes, and "Pluggie," a talking robotic fire hydrant. The Red Cross distributed informational literature to elementary and secondary schools for use in classes addressing home and family issues, and encouraged establishment of separate fire safety curricula. Smoke detectors and replacement batteries were distributed to the elderly and low-income families and public displays, billboards, and public service announcements were used to promote fire prevention awareness. (SB:ES; RB:HF,SD; PP:EC)

Organization:	**East St. Louis Volunteers**
Original Contact: Original Address:	Ms. M. Frances Nash Terrell East St. Louis Volunteers 144 N. 16th Street East St. Louis, Illinois 62205
Current Contact: Current Contact:	Unchanged Unchanged
Target Audience:	The general population, especially elementary school children and the elderly (ES, SA, GP)
Problem:	Lack of fire safety education has resulted in many fire-related deaths and injuries, especially among children and the elderly.
Program:	The Volunteers program featured distribution of smoke detectors and replacement batteries to the elderly and placement of detectors in schools. Fire department personnel gave fire safety talks and held fire drills at schools and gave fire safety presentations at senior citizens center meetings. The Volunteers used "Pluggie," a talking robotic fire hydrant, and poster contests to increase children's interest in fire safety. The Volunteers also sponsored the design of billboards, posters, and street banners and the placement of those banners throughout the city. (SB:ES; RB:HF,SD; PP:EC)

Organization:	**Madison County Firemen's Association**

Original Contact:	Capt. Dennis Henson
Original Address:	Madison County Firemen's Association
	9510 Collinsville Road
	Collinsville, Illinois 62234

Current Contact:	Unchanged
Current Address:	Unchanged

Target Audience: Elderly citizens and school children, especially third and fourth graders (ES, SA)

Problem: Most fire deaths have occurred in homes, yet very few homes ever have had fire safety inspections. In addition, children and the elderly have accounted for a larger proportion of fire deaths than any other age groups.

Program: The firemen's association instituted a county-wide junior fire marshal program, in which fire personnel, equipped with a slide/audio tape presentation and fire safety brochures, instructed third and fourth graders on how to conduct their own home fire safety inspections. The firefighters awarded junior fire marshal badges to students who returned completed inspection checklists. The program also employed "Pluggie," a talking robotic fire hydrant, to increase children's interest in fire safety. For senior citizens, the firefighters provided instruction on tire and burn prevention and supplied and assisted with installation and maintenance of smoke detectors. (SB:ES; RB:HF,SD; PP:EC)

Organization:	**Mokena Lions Club**

Original Contact:	Mr. Robert McGann
Original Address:	Mokena Lions Club
	11022 Hiawatha
	Mokena, Illinois 60448

Current Contact:	Unchanged
Current Address:	Unchanged

Target Audience: Juvenile firesetters and their families; children in preschool through ninth grade; and babysitters (PS, ES, SS, A)

Problem: About half of residential fires have been caused by children four to 12 years old. Children's curiosity about fire and lack of knowledge about fire safety have been two major causes of juvenile firesetting.

Program: The program's central focus was identification, education, and counseling of juvenile firesetters and their families. However, the program also included fire safety education programs for children in preschool through ninth grades and for babysitters. Fire safety literature, Smokey the Bear and Sparky the Fire Dog characters and A.L.E.R.T., a talking robotic fire hydrant, were used in schools and at fire prevention open houses to raise interest in tire safety. (SB:PS,ES; RB:HF; SF:AR)

INDIANA

Organization: **American Red Cross, Henry County Chapter**

Original Contact: Mr. James E. Nipp
Original Address: American Red Cross, Henry County Chapter
                           P. 0. Box 492
                           New Castle, Indiana 47362

Current Contact: Unchanged
Current Address: American Red Cross
                           2011 Bundy Avenue #4
                           New Castle, Indiana 47362-2912

Target Audience: Children in kindergarten through sixth grade, including deaf and blind children (ES, SP)

Problem: Many children have been killed or injured in house fires in Indiana, yet the average citizen
                           has little knowledge of what a fire is like and how to respond to one.

Program: In cooperation with local firefighters, the Red Cross developed an original video production
                           and an accompanying activity book that depict children learning about fire and responding to
                           fire emergencies. Firefighters and teachers worked together to introduce the videos and
                           activity books with informational talks in area elementary schools. The program featured
                           special sections addressing tire safety issues for deaf and blind children. (SB:ES; SF:HC)

Organization: **Greater Gary Chamber of Commerce**

Original Contact: Ms. Maxine J. Young
Original Address: Greater Gary Chamber of Commerce
                           504 Broadway
                           Suite 324
                           Gary, Indiana 46402

Current Contact: Unchanged
Current Address: Unchanged

Target audience: General urban population (GP)

Problem: Gary reported more arson fires in 1986 and 1987 than any other American city. Most of the
                           arson fires involved vacant and derelict buildings. Juvenile who were attracted to the
                           excitement of tire but not aware of its dangers were one major group of arsonists. Vandals
                           and mentally disturbed individuals also were responsible for a significant proportion of arson
                           fires.

Program: The chamber of commerce planned anti-arson and fire safety education programs using
                           videotaped materials, brochures, and a bureau of speakers on fire prevention presentations in
                           schools and at community events. A key feature of the program was a focus on basketball, a
                           very popular pastime among area residents, as a vehicle to promote fire safety; this element
                           of the program involved use of a "barnstorming" fire department basketball team to attract
                           crowds for fire safety presentations, and the assistance of local professional athletes to
                           publicize fire prevention activities. (SB:ES; SF:AR)

Organization: **Wayne Township 4-H Club**

Original Contact: Mrs. Patricia Eddleman
Original Address: Wayne Township 4-H Club
                        1220 South High School Road
                        Indianapolis, Indiana 46241

Current Contact: Unchanged
Current Address: Unchanged

Target Audience: Early adolescents in the sixth, seventh, and eighth grades (SS)

Problem: Wayne Township experienced a steep increase in arson fires and related injuries in 1987. Juveniles were responsible for setting many of the arson fires and for causing many accidental fires. Peer pressure was cited as a major factor encouraging youth to intentionally set fires, engage in careless behavior that can lead to fire, or ignore other youths' firesetting behavior.

Program: The program involved early adolescent youth in designing and presenting an anti-arson and fire safety teaching program appropriate for their own age group. In addition to providing fire safety information, the program featured information on handling peer pressure and using peer pressure to discourage firesetting behavior. Juveniles also contributed to the design of innovative videotaped presentations. One depicted youths coping with a home fire emergency, and a series of "Fire Court" stories depicted youth testifying about their actions in fire and arson situations and challenged students, as a "jury," to evaluate the youths' actions. The Four-H Club recruited local professional athletes to encourage the youths' anti-arson and fire prevention activities. (SB:ES)

Organization: **YMCA of Greater Fort Wayne**

Original Contact: Mr. David Brittenham
Original Address: YMCA of Greater Fort Wayne
                        226 E. Washington Boulevard
                        Fort Wayne, Indiana 46802

Current Contact: Unchanged
Current Address: Unchanged

Target Audience: The general public, especially children (GP, ES)

Problem: A lack of adequate knowledge among the general public on how to prevent and survive fires has increased citizens' risk of death or injury from fire.

Program: The YMCA sponsored the creation of an exhibit at a local firefighters' museum in an effort to increase people's interest in preventing fires and responding appropriately to fires. The exhibit, which depicted the damage that fires can do to the average home, consists of two rooms built and furnished to resemble rooms in an actual home. The rooms were burned and reconstructed in the museum. Coupled with the rooms is a video presentation in which a child demonstrates methods of escaping from a burning home. (RB:HF)

IOWA

Organization:	**Clear Lake Chamber of Commerce**

Original Contact: Original Address:	Mr. Richard Berrier Clear Lake Chamber of Commerce P. 0. Box 188 Clear Lake, Iowa 50428

Current Contact: Current Address:	Mr. Mark Schoeller Unchanged

Target Audience:	Elementary school children (ES)

Problem:	Citizens generally have lacked adequate fire prevention and survival education.

Program:	The project board of directors worked with local school officials to have one classroom equipped with displays for fire prevention and survival instruction. Local senior citizens volunteered to teach fire safety lessons in the classroom facility. Under the program, home inspections sheets and stickers with phone emergency number have been developed, printed, and distributed. (SB:ES)

Organization:	**Home Fire Safety Task Force of Eastern Iowa**

Original Contact: Original Address:	Ms. Florence Robertson Home Fire Safety Task Force of Eastern Iowa P. 0. Box 2776 Cedar Rapids, Iowa 52406

Current Contact: Current Address:	Mr. Ethan Sposton Unchanged

Target Audience:	Preschool and elementary school children, and low-income persons (PS, ES, SP)

Problem:	Persons with limited incomes and young children have not received adequate fire prevention and survival education.

Program:	The task force sponsored development of television and radio public service announcements and distribution of newsletters on fire prevention and survival. The task force also arranged fire safety presentations at local preschools and elementary schools. (SB:ES,PS; RB:HF)

Organization:	**Red Oak Rotary Club**

Original Contact:	Mr. David W. Hammer
Original Address:	Red Oak Rotary Club
	P. 0. Box 412
	Red Oak, Iowa 51566

Current Contact:	Unchanged
Current Address:	Unchanged

Target Audience: General Population (GP)

Problem: A lack of fire prevention and survival training among the general public has left the community vulnerable to fire-related deaths and injuries. Inadequate marking of street address numbers on houses hampers identification during emergencies. The need for increased maintenance for smoke detectors and the construction of fire exits for two-story buildings also was identified.

Program: The program involved attempts to mark unnumbered homes, installation of smoke detectors, distribution of support ladders for multiple-story homes without exits, fire and burn safety presentations, and distribution of informational brochures. (RB:HF,SD)

KENTUCKY

Organization: **Florence Lions Club**

Original Contact: Mr. Christopher Mehling
Original Address: Florence Lions Club
29 La Cresta Drive
Florence, Kentucky 41042

Current Contact: Unchanged
Current Address: Unchanged

Target Audience: Children and county officials (ES, A)

Problem: Residential fires have claimed the lives of many children, but local agencies and officials have lacked coordination in developing a united fire prevention effort.

Program: The Lions arranged meetings of local officials to discuss fire prevention activities, trained teachers in fire prevention education, and distributed tire-prevention educational materials in the local schools. (SB:ES; RB:HF)

Organization: **Pike County Chamber of Commerce**

Original Contact: Mr. Bryan Compton
Original Address: Pike County Chamber of Commerce
P.O. Box 897
Pikeville, Kentucky 41501

Current Contact: Ms. Dawn D. Hutchinson
Current Address: Unchanged

Target Audience: General population, elementary school students (GP, ES)

Problem: Arson had been identified as the major cause of severe increases in the fire loss rate in Pike County over the years.

Program: Programs using public speakers, videotapes, and brochures were presented at hi schools and at meetings of civic and service organizations. An existing "Keystone Firemen" fire education program for children was expanded into the fifth grade of area elementary schools. (RB:HF; SB:ES)

Organization:	**Portland Development Organization**

Original Contact: Ms. Molly Leonard
Original Address: Portland Development Organization
2915 Portland Avenue
Louisville, Kentucky 40212

Current Contact: Unchanged
Current Address: Unchanged

Targeted Audience: Elementary school children and the general population (ES,GP)

Problem: Children have been responsible for setting a large percentage of tires.

Program: The program featured development of a neighborhood arson watch system and arson-related resource library. The program also fostered arson-related play competitions for elementary, middle, and high school students as an educational tool and development of neighborhood youth groups to combat arson through use of peer pressure. (SB:ES; SF:AR)

LOUISIANA

Organization: **Kiwanis Club International**

Original Contact:
Original Address:

Mr. John M. DeFraites
Kiwanis Club International
Kiwanis Club of Houma, Inc.
P. O. Box 701
Houma, Louisiana 70361

Current Contact:
Current Address:

Mr. Thomas Campo
Unchanged

Target Audience:

The general population, with an emphasis on youth (GP, ES)

Problem:

Fire has claimed the lives of many area youths.

Program:

The Kiwanis helped implement a fire safety education curriculum in the schools, sought legislation to require public schools to implement fire safety curricula, and used the media to increase public awareness and support of the program. The program also produced an original video production on home fire safety. (SB:ES; RB:HF)

Organization: **The Knights of Columbus, St. Aiphonsus Council #3331**

Original Contact:
Original Address:

Mr. Lee Roy D. Campo
Knights of Columbus
St. Alphonsus Council #3331
6959 Airline Highway
Baton Rouge, Louisiana 70814

Current Contact:
Current Address:

Unchanged
Knights of Columbus
St. Alphonsus Council #3331
3025 Woodcrest Drive
Baton Rouge, Louisiana 70814

Target Audience:

The general public (GP)

Problem:

Improper electrical wiring, damaged electrical cords, and the misuse and improper storage of flammable liquids have caused many fire-related injuries and deaths and much property damage.

Program:

The program emphasized proper handling of electrical wiring and of flammable materials, encouraged the use of smoke detectors and fire extinguishers in the home, and stressed the importance of having a home fire escape plan. The program also involved a fire safety curriculum for the general public, and the use of broadcast public service announcements, billboards, and presentations in area malls to increase public fire safety knowledge. (RB:HF,SD)

MAINE

Organization: **Brown's Head Repertory Theatre**

Original Contact:	Ms. Kathy Huff
Original Address:	Brown's Head Repertory Theatre
	P. O. Box 248
	Dexter, Maine 04930

Current Contact:	Unchanged
Current Address:	Unchanged

Target Audience: Elementary school children (ES)

Problem: Children increasingly have become victims of fire-related incidents due to a lack of such fire prevention knowledge as an understanding about storage and use of flammable substances.

Program: The program resulted in the development of a 40-minute play to teach fire safety techniques to elementary school children during a tour of schools in area counties. The group produced a videotape of the play and support materials for the classroom. (SB:ES)

Organization: **Freeport Rotary Club**

Original Contact:	Mr. Paul Bennett, Jr.
Original Address:	Freeport Rotary Club
	P. O. Box 613
	Freeport, Maine 04032

Current Contact:	Mr. Henry Bird
Current Address:	Freeport Rotary Club
	P. O. Box 100
	Freeport, Maine 04032

Target Audience: Rural area general population (GP)

Problem: The fire department's response time is lengthy because Freeport is in a rural location and the fire department is a volunteer group, and fire prevention and survival education is therefore important. But many families were not practicing adequate fire prevention in the home, and many disconnected smoke detectors because of frequent false alerts. In addition, local volunteer fire departments had difficulty recruiting adequate numbers of volunteers.

Program: In cooperation with the Rotary Club, local fire department personnel developed two original videotaped programs and accompanying literature. One program addresses such home fire safety issues as smoke detector use and escape planning. The other, intended to enhance fire department volunteer recruitment activities, profiles the work of emergency services volunteers. The programs were shown in schools, at senior citizen group meetings, at a local factory, at a statewide fire personnel convention, and at several area fire departments. (RB:HF;SF:RF)

Organization:	**Presque Isle Lions Club**
Original Contact:	Mr. Joe Lloyd
Original Address:	Presque Isle Lions Club
	P. O. Box 442
	Presque Isle, Maine 04769
Current Contact:	Mr. Tom Deschaine
Current Address:	Presque Isle Lions Club
	67 Barton Street
	Presque Isle, Maine 04769
Target Audience:	Senior citizens (SA)
Problem:	Improper uses of smoking materials, oil furnaces, wood stoves, and electrical cooking appliances have caused many fire-related deaths among the elderly.
Program:	The Lions Club sponsored television public service announcements that featured local fire chiefs discussing home fire prevention and safety measures. The club also arranged home fire safety inspections and fire safety presentations to elderly groups. (PP:EC)

Organization:	**Thornton Heights Lions Club**
Original Contact:	Mr. Ralph Wink
Original Address:	Thornton Heights Lions Club
	c/o Portland Pipeline
	P. O. Box 2590
	South Portland, Maine 04106
Current Contact:	Unchanged
Current Address:	Unchanged
Target Audience:	Senior citizens (SA)
Problem:	Senior citizens have comprised the fastest-growing segment of the city's population. However, the city has not had sufficient resources to provide the elderly with fire prevention and survival education.
Program:	In cooperation with the Lions Club, the South Portland Fire Department developed an education/information program to be presented at housing developments for the elderly by fire department personnel. The program employs of audio/visual materials and workbooks suitable for presentation through television as well as in-person. Topic areas covered include planning evacuation and other actions to be taken in the event of fire, proper installation and care of smoke detectors, and fire hazards associated with cooking. (PP:EC)

MARYLAND

Organization: **Hollywood Optimist Club**

Original Contact: Mr. Charles Davis, Jr.
Original Address: Hollywood Optimist Club
     Hollywood, Maryland 20636

Current Contact: Unchanged
Current Address: Unchanged

Target Audience: Children ages six through 15 years old and the general public (ES, SS, GP)

Problem: The number of reported fires has been increasing each year, and there have been two deaths per year, on average, from fire.

Program: The club helped establish fire safety programs in public schools, private schools, and day care centers and set up fire prevention and survival courses for babysitters. The club also arranged for fire safety presentations in malls, museums, and fire department open houses, and for distribution of smoke detectors. The club used the news media to publicize program activities and fire safety issues. (SB:PS,ES; RB:HF,SD)

Organization: **Maryland Community Association for Education of Young Children**

Original Contact: Ms. Sallie Tinney
Original Address: Maryland Association for the Education of Young Children
     c/o The Center for Young Children
     University of Maryland
     College Park, Maryland 20742

Current Contact: Unchanged
Current Address: Maryland Community Association for Education of Young Children
     4601 Guilford Place
     College Park, Maryland 20740

Target Audience: Preschool children (PS)

Problem: Children under the age of six have accounted for the highest percentage of fire deaths of any age group in the state. Fear of firefighters and their equipment also has been a significant problem among children.

Program: In an effort to teach children how to respond to fires, the association developed an eight-day program for preschoolers that simulates real fire situations and produced literature that children can take home to their parents. The association also developed a curriculum guide and resource booklet and trained the teachers and day care providers to instruct children in fire safety. (SB:PS)

Organization: **Soroptimist International of Frederick County, Maryland**

Original Contact: Ms. Thelma Mullenix
Original Address: Soroptimist International of Frederick County, Maryland
                        498 Carrollton Drive
                        Frederick, Maryland 21701

Current Contact: Ms. Peggy Webb
Current Address: Soroptimist International of Frederick County, Maryland
                        6328 New Haven Court
                        Frederick, Maryland 21701

Target Audience: The elderly and the hearing impaired (SA, SP)

Problem: The elderly have lacked education on fire safety, and inexpensive smoke alarms suitable for the hearing impaired have not been available.

Program: The program featured development of a fire safety program for elderly residents, distribution of smoke detectors and batteries, and initiation of development of an inexpensive smoke detector for the hearing impaired. (PP:EC; SF:HC)

MASSACHUSETTS

Organization: **Justice Resource Institute, Inc.**

Original Contact: Ms. Margaret Friedman
Original Address: Justice Resource Institute, Inc.
                     132 Boylston Street
                     Boston, Massachusetts 02116

Current Contact: Mr. Joel Kershner
Current Address: Unchanged

Target Audience: The general public, with emphasis on elderly and minority groups (GP, SA, SP)

Problem: A lack of fire prevention awareness, caused by a general sense of apathy, made community residents vulnerable to fire.

Program: The institute developed a presentation on arson and fire prevention for the Southeast Asian population, sponsored public service announcements on local radio stations, and distributed smoke detectors to elderly and low-income individuals. The group also initiated formation of neighborhood arson watch groups and produced original community training manuals for those programs. (RF:HF,SD; PP:EC; SF:AR)

Organization: **Wenham-Hamilton Lions Club**

Original Contact: Mr. Glenn Herrick
Original Address: Wenham-Hamilton Lions Club
                     28 Mayflower Drive
                     Wenham, Massachusetts 01984

Current Contact: Mr. Robert W. Moroney
Current Address: Hamilton-Wenham Lions Club
                     11 Howard Street
                     Wenham, Massachusetts 01984

Target Audience: The general public (GP)

Problem: The many large wooden structures in the community have caused concerns about fire safety, but there has been a lack of qualified personnel to install sprinkler systems in vulnerable buildings.

Program: The club arranged installation of sprinkler systems in two wood-frame units to train personnel to install such systems in other buildings. In addition, the club produced a slide presentation and distributed literature throughout the community to make the public more aware of the need for sprinkler systems in residential homes. (RB:RS)

MISSISSIPPI

Organization: **Batesville Junior Auxiliary**

Original Contact: Ms. Betty Ray Crume
Original Address: P. O. Box 885
                    Batesville, Mississippi 38606

Current Contact: Unchanged
Current Address: Unchanged

Target Audience: Low-income residents and elementary school children (SP, ES)

Problem: Crowded living quarters, poor wiring, and a lack of fire safety knowledge have caused a great number of fires among low-income residents.

Program: The auxiliary club conducted a survey to evaluate the problem and target those individuals most in need of help. The club distributed smoke detectors to low-income residents who agreed to attend a fire safety information session and arranged featured fire safety demonstrations before community organizations and fire safety booths at local public events. The club also helped establish a fire safety curriculum in the elementary schools. (SB:ES; RB:HF,SD)

Organization: **George County Jaycees**

Original Contact: Mr. Ricky Churchwell
Original Address: George County Jaycees
                    P. O. Box 733
                    Lucedale, Mississippi 39452

Current Contact: Unchanged
Current Address: P. O. Box 405
                    Lucedale, Mississippi 39452

Target Audience: General public, with emphasis on third and fourth grade school children and the elderly (GP, ES,SA)

Problem: A general lack of fire safety knowledge has contributed to many fire hazards, including lack of home fire escape planning, improper use of flammable fuels, unsafe cooking practices.

Program: The Jaycees helped establish a fire safety curriculum in elementary schools. In cooperation with fire department personnel, the Jaycees actively participated in National *Fire* Prevention Week by showing fire prevention films, distributing brochures, and demonstrating the use of fire safety equipment. In addition, the Jaycees arranged for fire personnel to install smoke detectors in homes of senior citizens. (SB:ES; RB:HF,SD; PP:EC)

Organization:	**Greater Meridian Jaycees**
Original Contact:	Mr. Archie Anderson
Original Address:	P.O. Box 1743
	Meridian, Mississippi 39301
Current Contact:	Unchanged
Current Address:	Unchanged
Target Audience:	General public (GP)
Problem:	A lack of fire safety and prevention education has left community residents vulnerable to fire.
Program:	The Jaycees, in cooperation with local government officials and fire departments, provided fire education programs for schools, civic clubs, nursing homes, hospitals, and local industries. Materials used in the programs included videotapes, literature, and a mobile trailer unit. (RB:HF)

MISSOURI

Organization: **Cherokee Parent-Teacher Association**

Original Contact: Mr. Dennis Greeson
Original Address: Cherokee Parent-Teacher Association
                     420 East Plainville Road
                     Springfield, Missouri 65807

Current Contact: Unchanged
Current Address: Cherokee Parent-Teacher Association
                     5002 Coach Road
                     Battlefield, Missouri 65619

Target Audience: Children under 14 years of age (ES)

Problem: Youths have been responsible for a high percentage of arson fires and have lacked adequate fire prevention and safety education.

Program: The PTA developed a program based on a fictitious fire safety advocator, "Captain Nomex," who gave presentations at local schools that included slide shows, group demonstrations, and distribution of Captain Nomex workbooks. (SBES)

Organization: **Pickering Lions Club**

Original Contact: Mr. Ross Johnson, Jr.
Original Address: Pickering Lions Club
                     Pickering, Missouri 64476

Current Contact: Unchanged
Current Address: Unchanged

Target Audience: Third- and fourth-grade children, the elderly, and civic groups (ES, SA, GP)

Problem: Pickering is a rural farm community with only a small, poorly equipped volunteer fire department and without a fire water distribution system. Most residents have not had adequate fire prevention and survival education, and youths and the elderly have been especially vulnerable to deaths and injuries from accidental fires.

Program: For the children, a fire safety videotape presentation on how to prevent and escape home fires was created and presented in several schools and at public fairs. Smoke detectors, along with follow-up information on proper battery maintenance, were distributed to the elderly and low-income populations. (SB:ES; PP:EC, SF:RF)

Organization:	**West St. Louis County Chamber of Commerce**

Original Contact:
Original Address:

Ms. Jenny Huggins
West St. Louis County Chamber of Commerce
910 Clayton Road, Suite 310
St. Louis, Missouri 63011

Current Contact:
Current Address:

Ms. Marian Silvernail
West St. Louis County Chamber of Commerce
304 Temple Avenue, Suite 107
Ballwin, Missouri 63011

Target Audience:

The general population, elementary school children, and the elderly (GP, ES, SA)

Problem:

Young children and the elderly have experienced high fire injury and fatality rates, and youths have been responsible for starting many fires. A general lack of fire safety education within the community has contributed to fire problems.

Program:

The program involved fire prevention and survival presentations at local schools, installation of smoke detectors in homes of the elderly, distribution of 911 bumper stickers, and presentation of an information booth at a local business fair. A local store offered discounts on smoke detectors and house numbers to support the program. (SB:ES; RB:HF,SD; PP:EC)

Organization:	**Zonta International, Cape Girardeau Club**

Original Contact:
Original Address:

Ms. Cynthia A. Pitman
Zonta International, Cape Girardeau Club
c/o Cape Girardeau Fire Department
1 South Sprigg Street
Cape Girardeau, Missouri 63701

Current Contact:
Current Address:

Ms. Dorothy Gilbert
Unchanged

Target Audience:

Elementary school children, the elderly, and the general public (ES, SA, GP)

Problem:

Children and the elderly experienced high rates of fire-related injuries and deaths, and the general public lacked education on fire safety, home escape planning, and smoke detector maintenance.

Program:

"Stop, Drop, and Roll" and "Get Low and Go" fire survival presentations were introduced in local grade schools. Smoke detectors were installed in elderly individuals' homes, along with home visits to discuss maintenance of detectors. Program also included distribution of a newsletter on fire safety tips and the broadcasting of public service announcements. (SB:ES; PP:EC; RB:HF)

NEW YORK

Organization:	**Greater Troy Chamber of Commerce**

Original Contact: Mr. John O'Connor
Original Address: Greater Troy Chamber of Commerce
                             3251 River Street
                             Troy, New York 12180

Current Contact: Unchanged
Current Address: Unchanged

Target Audience: Kindergarten through sixth grade children, the elderly, and the general public (ES, SA, GP)

Problem: There was a great need to expand fire safety education to all age groups. In addition, many senior citizens have died as a result of preventable kitchen fires.

Program: The chamber of commerce helped to establish fire safety curriculum programs in schools and brought fire safety puppet shows to the schools. Informational programs on fire safety in the home and at work were held at meetings of senior citizen groups and local service- and business-oriented organizations. Fire safety-related display booths were set up at fairs and community events. The program also used public service announcements and posters on public transportation vehicles to publicize use of smoke detectors and other fire safety issues. (SB:ES; RB:HF,SD; PP:EC)

NORTH CAROLINA

Organization:	**Monroe Avenue Parent-Teacher Organization**

Original Contact: Mr. Jim Covington
Original Address: Monroe Avenue Parent-Teacher Organization
400 Monroe Avenue
Hamlet, North Carolina 28345

Current Contact: Mr. Calvin White
Original Address: Unchanged

Target Audience: Elementary school children (ES)

Problem: Unsupervised children have been responsible for many local fires.

Program: The organization developed a school information program to encourage children to adopt fire safety habits and teach those habits to their parents. Each grade was given a specific task and related activities. For example, fourth graders were assigned to install smoke detectors, while third graders became honorary members of the fire department. (SB:ES)

NORTH DAKOTA

organization: **North Dakota Safety Council, Inc.--Bismarck Residential Sprinkler Program**

Original Contact: Mr. Brian Larson
Original Address: North Dakota Safety Council, Inc.
     4023 State Street
     Bismarck, North Dakota 58501

Current Contact: Ms. Cheri Hust
Current Address: North Dakota Safety Council, Inc.
     111 N. 6th Street
     Bismarck, ND 58501

Target Audience: The general population, especially rural residents and those without residential sprinkler systems (GP)

Problem: More than 50 fires and more than $200,000 in financial losses due to fires occurred annually in the area. Causes include solid fuel-burning appliances, gas appliances, combustibles placed too close to heat sources, improperly installed equipment, fireworks, and lightning. A lack of knowledge about escaping from a fire have contributed to the problem.

Program: The Safety Council built a mobile unit demonstrating the proper installation and operation of a residential sprinkler system; displayed the mobile unit at building trade shows and other appropriate events; developed and distributed literature on residential sprinklers for use in fire education seminars; and installed residential sprinkler systems in several buildings. (RB:HF,RS; SE:RF)

Organization: **North Dakota Safety Council, Inc.--Oliver County Rural Fire Prevention Program**

Original Contact: Mr. Brian Larson
Original Address: North Dakota Safety Council, Inc.
     4023 State Street
     Bismarck, North Dakota 58501

Current Contact: Ms. Cheri Hust
Current Address: North Dakota Safety Council, Inc.
     111 N. 6th Street
     Bismarck, ND 58501

Target Audience: Rural residents, especially families living on farms (GP)

Problem: The number of rural site fires resulting from improper storage of flammable materials and refueling of equipment increased dramatically.

Program: The program involved construction and display of a mobile fire safety demonstration unit that consisted of a farm diorama used to demonstrate fire hazards and prevention at rural residences. The unit was used to increase fire prevention awareness regarding use of fire extinguishers on farm equipment and early warning devices in farm buildings, safety tips on open burning, and development of exit routes in the event of a home fire. Audiovisual aids, pamphlets, and brochures also were distributed, and a media campaign was conducted (RB:HF; SF:RF)

Organization:	**Williston Jaycees/North Dakota Safety Council, Inc.**

Original Contact:	Ms. April Nygard
Original Address:	Williston Jaycees
	P. O. Box 1314
	Williston, North Dakota 58801

Current Contact:	Ms. Cheri Hust
Current Address:	North Dakota Safety Council, Inc.
	111 N. 6th Street
	Bismarck, North Dakota 58501

Target Audience:	Elementary school children and the general adult population (ES, GP)

Problem:	Nearly three-quarters of all structural fires have occurred in mobile homes. In addition, widespread use of such alternative heating devices as fireplaces, wood stoves, electric space heaters, kerosene heaters, and propane heaters posed a significant fire threat.

Program:	The program employed a mobile home to demonstrate typical mobile home fire hazards. The mobile home was equipped with a smoke machine for use in home fire escape drills. The program also equipped a trailer with several alternative heating devices to demonstrate proper installation and use of the devices and conducted demonstrations at local community events. Instructional presentations that made use of brochures, films, and videotapes were made at schools, civic group meetings, and community events. In addition, smoke detectors and fire extinguishers were distributed. (SB:ES; RB:HF,SD)

OKLAHOMA

Organization: **Community Action Agency of Oklahoma City**

Original Contact: Mr. James E. Sconzo
Original Address: Community Action Agency of Oklahoma City
                        1900 Northwest Tenth Street
                        Oklahoma City, Oklahoma 73106

Current Contact: Unchanged
Current Address: Unchanged

Target Audience: Preschool children attending day care centers in high risk areas of Oklahoma City (PS)

Problem: Children under the age of six have not been trained in fire and burn prevention, fire survival, and recognition of the firefighter as a helper of children.

Program: The program involved enhancing a locally developed preschool fire safety education program by developing the existing curriculum into a professional product; purchasing and producing video tapes to support the fire safety lessons; training volunteers as fire safety instructors; and providing parent tire safety follow-up kits. (SB:PS)

Organization: **Poteau Kiwanis** Club

Original Contact: Mr. Jim Smith
Original Address: Poteau Kiwanis Club
                        P. O. Box 512
                        Poteau, Oklahoma 74953

Current Contact: Unchanged
Current Address: Unchanged

Target Audience: High school students and members of the Future Farmers of America (FFA) (SS)

Problem: Debris fires in forest and grassland areas have blackened large areas and caused injuries to firefighters and landowners. The rural community has not been educated in proper controlled burning techniques or available resources for help in suppressing the fires.

Program: The program involved teaching the correct method for burning debris, responding to wildland fires, and teaching correct behavior for responding to clothing fires and burn accidents. The program developed a fire safety education program for FFA students, including slide/tape programs and instructional materials which were provided to area schools; recruited and trained FFA teachers and other community volunteers to teach the program in the classroom; field-tested the program; and increased public awareness of the program by distributing newsletters and mailings. (SB:ES; SF:WL)

SOUTH CAROLINA

Organization: **Aiken County Independent Insurance Agents**

Original Contact: Mr. David Derrick
Original Address: Aiken County Independent Insurance Agents
                      Volunteer Fire Prevention Project
                      129 Park Avenue
                      Aiken, South Carolina 29801

Current Contact: Unchanged
Current Address: Unchanged

Target Audience: Elementary school children and the elderly (ES, SA)

Problem: Most residential fires are attributable to a lack of fire safety education and carelessness have among elementary schoolchildren and the elderly.

Program: The program featured implementation of a fire safety curriculum program in 20 schools. The group also worked with the Public Safety Department to build a fire safety trailer and install smoke detectors in homes of the elderly. The group used the news media to supplement the program. (SB:ES; RB:SD; PP:EC)

Organization: **Blue Ridge Association of Insurance Women, Inc./Clemson Jaycees**

Original Contact: Ms. Linda Rice
Original Address: Blue Ridge Association of Insurance Women, Inc.
                      c/o Clemson University
                      204 Sikes Hall
                      Clemson, South Carolina 29634

Current Contact: Mr. Jack Abraham
Current Address: Clemson Fire Department
                      Perimeter Road
                      Clemson, South Carolina 29634-5510

Target Audience: College students and the elderly (SS, SA)

Problem: Many fires are attributable to a lack of fire safety education and a lack of smoke detectors in homes. In addition, many buildings do not conform with proper building practices and safety designs.

Program: The association coordinated a general public fire safety program that included the distribution of smoke detectors to elderly residents. The group also conducted a survey of Clemson University students to determine their level of fire prevention knowledge and develop a fire safety curriculum program to meet the needs of college students. (RB:SD; PP:EC)

Organization:	**Clarendon Business and Professional Women's Club**

Original Contact:	Ms. Christine Brewer
Original Address:	Clarendon Business and Professional Women's Club
	Route 3, Box 840
	Manning, South Carolina 29102

| Current Contact: | Unchanged |
| Current Address: | Unchanged |

Target Audience: The general population, with an emphasis on elementary school children (GP, ES)

Problem: Effective educational programs for children have been lacking. There also has been a lack of information about the nature of arson and an absence of smoke detectors in residences.

Program: The program featured implementation of a planned, consistent public fire education program, including arson education, in schools in the county. The program also involved distribution of smoke detectors to high risk populations and assistance with installation and home fire safety inspections. (SB:ES; RB:HF,SD; SF:AR)

Organization:	**Russell Massey and Co.**

Original Contact:	Mr. Charles Dorton
Original Address:	Independent Insurance Agents of Columbia
	P. 0. Box 5801
	Columbia, South Carolina 29250

Current Contact:	Unchanged
Current Address:	Russell Massey and Co.
	P. 0. Box 7871
	Columbia, South Carolina 29202

Target Audience: The general population (GP)

Problem: The fire death rate has been high, and elderly residents and low-income families have suffered most from fires.

Program: The group helped develop a fire safety trailer and an accompanying videotape showing how fire safety techniques, and how to suppress and escape fires. The group also distributed smoke detectors and used the news media to publicize fire safety issues and program activities. (RB:HF,SD)

Organization: **Rotary Club of Bennettsville**

Original Contact: Mr. Michael O'Tuel
Original Address: Rotary Club of Bennettsville
                        Box 444
                        Bennettsville, South Carolina 29512

Current Contact: Mr. Edward Anderson
Current Address: Unchanged

Target Audience: The general population (GP)

Problem: Improper use of alternative heaters has caused numerous fires; kitchen fires have threatened many lives; and more chimney fires have occurred due to lack of maintenance and lack of annual inspections.

Program: Fire personnel taught fire prevention and survival to school children in the kindergarten through third grade. Community workshops held during the entire year, culminated in activities during a fire prevention week. The group used the news media to publicize program activities. (RB:HF)

SOUTH DAKOTA

Organization: **Community School Organization, Ethan Fire Prevention Project**

Original Contact: Ms. Nancy Schoenfelder
Original Address: Community School Organization
Ethan Fire Prevention Project
Box 8
Ethan, South Dakota 57334

Current Contact: Unchanged
Current Address: Unchanged

Target Audience: Elementary school children, junior and senior high school students, and the general public (ES, SS, GP)

Problem: Fire injuries and deaths have increased due to a lack of escape plans and fire drills, lack of smoke detector maintenance, and lack of knowledge of directions to and street numbers of rural homes. Large numbers of children left alone or in the care of other children during the day create an additional fire safety concern.

Program: The program consisted of a fire safety information program presented to children in local schools, through the media, and in seminar presentations. Program administrators also trained volunteers; tested awareness and knowledge of fire safety; and educated youngsters and adults on safe cooking techniques, home exit plans, and the use of fire extinguishers. In addition, home escape ladders were distributed, and a school poster contest was held. (SB:ES)

Organization: **McKennan Hospital Burn Program**

Original Contact: Ms. Maria DeVaney
Original Address: McKennan Hospital Burn Program
800 East 21st Street
Sioux Falls, South Dakota 57117-5045

Current Contact: Unchanged
Current Address: Unchanged

Target Audience: Young children, parents of infants, and farm and other rural residents (GP)

Problem: There has been a high incidence of flammable liquid fires involving charcoal lighter fluid, gasoline, and kerosene around homes and on farms.

Program: To reduce the incidence of liquid and fire burns, program managers developed a locally produced video to educate youngsters and their parents about rural and farm fire prevention, produced and operated a puppet show, and developed and distributed education information. The program also involved and the development of a public information campaign using billboards. (RB:HF; SF:RF)

TENNESSEE

Organization:	**Breakfast Rotary Club**
Original Contact: Original Address:	Mr. David L. McCoy Breakfast Rotary Club P. O. Box 164 Oak Ridge, Tennessee 37831
Current Contact: Current Address:	Unchanged Unchanged
Target Audience:	Elementary school children and the elderly (ES, SA)
Problem:	Elementary school children and the elderly generally lack adequate fire safety and survival education.
Program:	The club helped continue fire safety curriculum programs in all public schools and expanded the programs into private schools. The program sponsored and conducted training sessions for community volunteers and teachers who, in turn, provided instruction to the respective target audiences. The project also continued the development and presentation of fire safety survival instruction for the elderly. (SB:ES; PP:EC)

Organization:	**Security Assistance For the Elderly (S. A. F. E.), Inc.**
Original Contact: Original Address:	Mr. Robert Horton S. A. F. E., Inc. Room 218 311 23rd Avenue, N Nashville, Tennessee 37203
Current Contact: Current Address:	Unchanged Unchanged
Target Audience:	The elderly and their caretakers (SA)
Problem:	Fire safety awareness has been lacking among senior citizens. In addition, a fire safety hazard has been created by residents' installation of bars over windows and double-locks on doors because of their fear of crime.
Program:	Volunteer instructors were trained in fire awareness, fire prevention, and escape planning. Those volunteers then trained other volunteers using visual aids developed for the original training program. Volunteers inspected senior citizens' homes for fire safety, installed smoke detectors, and conducted fire safety awareness classes at senior citizens centers, churches, and neighborhood association meetings. The program also provided training to staff at local hospitals and the Board of Health so that they, in turn, could instruct their elderly patients on fire safety, and SA.F.E. board members appeared on local television and radio programs to discuss the problems created when the elderly try to protect themselves from crime and inadvertently create fire traps in their homes. (RB:HF,SD; PP:EC)

TEXAS

Organization:	**Noon Optimist Club of Galveston**
Original Contact:	Dr. Duane Ullman
Original Address:	Noon Optimist Club of Galveston
	P. O. Box 567
	Galveston, Texas 77550
Current Contact:	Mr. Donald Poole
Current Address:	Unchanged

Target Audience: The elderly (SA)

Problem: The elderly generally have had very little formal training in fire and burn prevention and hence are susceptible to fire-related injury and death. A high percentage of area residential fires have occurred in homes that do not have smoke detectors.

Program: A three-part program involved teaching the elderly how to install and maintain smoke detectors, how to recognize home fire hazards and survive a residential fire, and how to work as a team in establishing fire brigades and developing evacuation plans in retirement centers. Program volunteers installed smoke detectors and developed fire safety brochures, public service announcements, and a videotape explaining the importance of installing and maintaining smoke detectors and practicing good fire safety habits. (RB:HF,SD; PP:EC)

Organizatlon:	**Pettus Rotary Club**
Original Contact:	Mr. Lee Willis
Original Address:	Pettus Rotary Club
	P. O. Box 115
	Tulleta, Texas 78146
Current Contact:	Mr. Clinton Bagwell
Current Address:	Unchanged

Target Audience: Elementary school children and the elderly (ES, SA)

Problem: Most fires are in homes and rural areas.

Program: The program provided fire prevention and education programs, including videotapes, to elementary school children and to the elderly; supported existing community fire safety programs; distributed smoke detectors and fire extinguishers to area residents; and developed and distributed brochures *on* fire reporting and heating safety. (SBES; RB:HF,SD; PP:EC)

UTAH

Organization:	**Handy Helpers 4-H Club**
Original Contact: Original Address:	Ms. Beverly Evans Handy Helpers 4-H Club Star Route Box 36 Altamont, Utah 84001
Current Contact: Current Address:	Unchanged Unchanged
Target Audience:	The general public, with an emphasis on elementary school students and the elderly (GP, ES, SA)
Problem:	Continuing fire hazards include active oil fields and an oil refinery operating near the community; alternative heating devices necessitated by extreme changes in climate; vacant lots containing trash, debris, and combustible materials near residential and business areas and in gullies. Fire losses have totalled more than $500,000 annually, with weed burning comprising the most serious fire hazard.
Program:	Volunteers and recruits implemented a community clean-up program; installed smoke detectors in the homes of senior citizens and those needing public assistance; conducted educational programs in schools; mounted monthly media campaigns with varying themes through print, television, and radio; and involved businesses and civic organizations in fire prevention activities. (SB:ES; RB:HF,SD; PP:EC)
Organization:	**Pleasant Grove Fire Ladies Auxiliary**
Original Contact: Original Address:	Ms. Carolyn Smith Pleasant Grove Fire Ladies Auxiliary 170 North 900 East Pleasant Grove, Utah 84062
Current Contact: Current Address:	Unchanged Unchanged
Target Audience:	The general population, civic organizations, elementary and high school students, babysitters, and the elderly (GP, ES, SS, SA)
Problem:	Lack of fire education for school age children has left them vulnerable to fire-related death and injury. Carelessness in managing rubbish fires and in using fireworks, improper installation of wood-burning stoves and chimneys, poor disposal of ashes, lack of smoke detectors and fire extinguishers, and lack of inspection of businesses for fire safety requirements also have created fire risks.
Program:	The programs focused on education regarding basic fire safety procedures, including the safe storage of hazardous and flammable substances. The group conducted fire education through displays at civic events and through programs and presentations in local schools, churches, and businesses and before civic groups, clubs, and other organizations. In addition, volunteers distributed and installed smoke detectors and fire extinguishers to elderly residents, inspected nursing homes and day care centers, and conducted a media awareness and education campaign. (SB:ES; RB:HF,SD; PP:EC)

VERMONT

Organization: **Fairbanks Museum and Planetarium**

Original Contact: Mr. Howard Reed
Original Address: Fairbanks Museum and Planetarium
Main and Prospect Streets
St. Johnsbury, Vermont 05819

Current Contact: Unchanged
Current Address: Unchanged

Target Audience: Elementary school students (ES)

Problem: Lack of adequate fire prevention education for elementary school children has left them vulnerable to fire-related death and injury.

Program: The museum developed materials on fire safety to be used in schools and with other groups of young people. The museum program sought to teach the "why's and how's" as well as the "do's and don't's" of fire prevention. The program used a video program that taught teachers how to instruct their students. (SB:ES)

Organization: **Fanny Allen Hospital**

Original Contact: Ms. Amy Raiser
Original Address: Fanny Allen Hospital
101 College Parkway
Winooski, Vermont 05404

Current Contact: Unchanged
Current Address: Fanny Allen Hospital
Medical Center
125 College Parkway
Colchester, VT 05446

Target Audience: The elderly and their caregivers (SA)

Problem: The elderly have been especially vulnerable to fires and burns.

Program: The hospital developed an education campaign to train caregivers in tire prevention and survival techniques, and media a campaign of radio public service announcements for the elderly. The hospital also distributed brochures and produced a 15-minute video highlighting tire survival skills. (PP:EC)

Organization: **Historic Windsor, Inc., Preservation Institute for the Building Crafts**

Original Contact: Ms. Judy Hayward
Original Address: Historic Windsor, Inc., Preservation Institute for the Building Crafts
                       P. O. Box 1777
                       Windsor, Vermont 05089

Current Contact: Unchanged
Current Address: Unchanged

Target Audience: Contractors, developers, restoration craftsmen (SP)

Problem: Conflicts between historic preservation and fire safety codes have left historical structures susceptible to fire.

Program: The group created a training program for builders and developers on historic restoration and fire safety to save structures of significant historical interest from damage and destruction. The program involved a series of workshops for anyone who was interested, and the group surveyed the participants to gain insight for further development of their program. (CP)

VIRGINIA

Organization: **Hampton University School of Education**

Original Contact: Dr. Mary T. Christian
Original Address: Hampton University School of Education
           Hampton, Virginia 23668

Current Contact: Unchanged
Current Contact: Unchanged

Target Audience: The general population, with a special emphasis on school children and juvenile firesetters (GP, ES)

Problem: Many fires are attributable to a general lack of awareness about fire safety, insufficient staff to teach fire safety education in the schools, little attention to the problem of juvenile firesetters, and a lack of city fire department resources for producing professional media messages on fire safety.

Program: The Hampton Fire Department trained Hampton University education department volunteers to help present lectures to school-age and civic groups in the community. Hampton University psychology department volunteers were trained to assist in the interviewing and education intervention of identified juvenile firesetters and were charged with developing an accurate data base for the juvenile firesetter program. The university mass media department provided interns to develop a complete public information campaign for fire prevention, including audio and video public service announcements used for national fire prevention week. (SB:ES; RB:HF; SF:AR)

Organization: **Total Action Against Poverty (TAP), Inc.**

Original Contact: Ms. Martha Ogden
Original Address: Total Action Against Poverty (TAP), Inc.
           P. O. Box 2868
           Roanoke, Virginia 24001

Current Contact: Mr. Correlli Rasheed
Current Address: Unchanged

Target Audience: The elderly and the handicapped, and small children living in low-income housing are at a high risk for fire (SA, SP, ES)

Problem: No early-warning detection systems in the homes of many low-income residents greatly increases their susceptibility to fire.

Program: The group distributed and installed smoke detectors and tire extinguishers in low-income homes with an identified need; distributed brochures and flyers to clients; spoke to neighborhood organizations and community groups on fire safety; appeared on local television and radio shows; participated in the annual tire prevention carnival; and conducted a puppet show on fire safety in a local elementary school. Program efforts were coordinated with existing educational programs of the city fire department. (SB:ES; RB:HF,SD)

WEST VIRGINIA

Organization:	**Beckley-Raleigh Chamber of Commerce**

Original Contact:	Mr. William A. Wilbur
Original Address:	Beckley-Raleigh Chamber of Commerce
	106 McCreary St., P. O. Box 1798
	Beckley, West Virginia 23302-1798

Current Contact:	Unchanged
Current Address:	Unchanged

Target Audience: Elementary school and junior high school students and teachers, civic organizations, and the elderly (ES, SS, GP, SA)

Problem: A lack of fire prevention education and awareness and a lack of smoke detectors has left area homes vulnerable to fires.

Program: The program involved the presentation of fire safety curriculum programs and the placement of curriculum books in area elementary schools; a "Fire House Mouse" computer exhibit shown at civic and other community events that demonstrated fire safety practices; a Science of Fire Safety presentation in schools and at a local museum; mall demonstrations and literature distribution; use of "Pluggie," a talking robotic fire hydrant; and fire safety presentations in nursing homes, hospitals, and day care centers. (SB:ES; PP:EC)

Organization:	**Kiwanis Club of Keyser**

Original Contact:	Mr. David C. Harman
Original Address:	Kiwanis Club of Keyser
	136 James Street
	Keyser, West Virginia 26726

Current Contact:	Unchanged
Current Address:	Unchanged

Target Audience: The general population, workers in public buildings such as churches and hospitals; elementary, junior high, and high school students; senior citizens; the hearing impaired; and mobile home owners (GP, ES, SS, SA, SP)

Problem: Among the fire-related problems in the area have been lack of smoke detectors in homes, automatic alarm malfunctions, and negligence and carelessness in the home. Alternative heaters and mobile home fires also have been identified as problems.

Program: Facilitators conducted fire safety curriculum programs, including film, print, and oral presentations in the schools; provided technical assistance and education of owners of public buildings; and distributed and installed smoke detectors, including detectors for the hearing impaired. (SB:ES; RB:HF,SD; SF:HC)

APPENDIX 2

NCVFPP TARGET AUDIENCE CATEGORIES

NCVFPP TARGET AUDIENCE CATEGORIES

Preschool--Ages 2-5 (PS)

Primary/Elementary School-&es 6-12 (ES)

Ballwin, Missouri; West St. Louis County Chamber of Commerce 80
Battlefield, Missouri; Cherokee PTA 79
Cape Girardeau, Missouri; Zonta International, Cape Girardeau Club 80
Kirksville, Missouri, Kirksville Noon Lions Club 17
Pickering, Missouri; Pickering Lions Club 79
Troy, New York; Greater Troy Chamber of Commerce 81
Beaufort, North Carolina; Carteret Community Action, Inc. 21
Hamlet, North Carolina; Monroe Avenue PTO 83
Jacksonville, North Carolina; Onslow County Agricultural Extension 22
Williston, North Dakota; Williston Jaycees/North Dakota Safety Council, Inc. 86
Tulsa, Oklahoma; American Red Cross, Tulsa Area Chapter 25
Harrisburg, Pennsylvania, Museum of Scientific Discovery 29
Aiken, South Carolina; Aiken County Independent Insurance Agents 89
Manning, South Carolina; Clarendon Business and Professional Women's Club 90
Ethan, South Dakota; Community School Organization 93
Huron, South Dakota; Retired Senior Volunteer Program of Beadle County 33
Marion, South Dakota; Marion Area Jaycees/Marion Area Senior Citizens Center, Inc. 32
Collierville, Tennessee; Chamber of Commerce 33
Oak Ridge, Tennessee; Breakfast Rotary Club 95
Beaumont, Texas; American Red Cross, Beaumont Chapter 35
Tulleta Texas; Pettus Rotary Club 97
Altamont, Utah; Handy Helpers 4-H Club 99
Ogden, Utah; Ogden YWCA 37
Pleasant Grove, Utah; Pleasant Grove Fire Ladies Auxiliary 99
St. Johnsbury, Vermont; Fairbanks Museum and Planetarium 101
Hampton, Virginia; Hampton University School of Education 103
Loudoun County, Virginia; Loudoun Business and Professional Women's Club 39
Roanoke, Virginia; Total Action Against Poverty, Inc. 103
Beckley, West Virginia; Beckley-Raleigh Chamber of Commerce 105
Charleston, West Virginia; Charleston Chamber of Commerce 41
Keyser, West Virginia; Kiwanis Club of Keyser 105

Secondary School-Ages 13-20 (SS)

Covina, California; Covina Women's Club 49
Rancho Cucamonga, California; We Tip, Inc., War on Arson 8
Miami, Florida; Urban League of Greater Miami, Inc. 55
West Point, Georgia; Professional Secretaries International 58
Mokena, Illinois; Mokena Lions Club 62
Peoria, Illinois; American Red Cross, Central Illinois Chapter 61
Indianapolis, Indiana; Wayne Township 4-H Club 64
Hollywood, Maryland; Hollywood Optimist Club 73
Kirksville, Missouri; Kirksville Noon Lions Club 17
Poteau, Oklahoma; Poteau Kiwanis Club 87
Clemson, South Carolina; Blue Ridge Association of Insurance Women/Clemson Jaycees 89
Ethan, South Dakota; Community School Organization 93
Pleasant Grove, Utah, Pleasant Grove Fire Ladies Auxiliary 99
Loudoun County, Virginia; Loudoun Business and Professional Women's Club 39
Beckley, West Virginia; Beckley-Raleigh Chamber of Commerce 105
Keyser, West Virginia; Kiwanis Club of Keyser 105

NCVFPP TARGET AUDIENCE CATEGORIES

Adult-Ages 21-64 (A)

Gadsden, Alabama; Civitans 45
Bethel, Alaska; Yukon-Kuskokwim Health Corporation 48
Sacramento, California; Fire Prevention Council 7
Mokena, Illinois; Mokena Lions Club 62
Florence, Kentucky; Florence Lions Club 67
Springfield, Massachusetts; Safety Council of Western Massachusetts 15
Albion, New York; Orleans/Genesee Rural Prevention Corporation, Inc. 19
Marion, South Dakota; Marion Area Jaycees/Marion Area Senior Citizens Center, Inc. ... 32
Ogden, Utah; Ogden YWCA 37
Loudoun County, Virginia; Loudoun Business and Professional Women's Club 39

Senior Adult-Ages 65+ (SA)

Birmingham, Alabama; University of Alabama at Birmingham Burn Center 5
Huntsville, Alabama; Exchange Club 45
Orange Beach, Alabama; Chamber of Commerce 5
Sacramento, California; Fire Prevention Council 7
Seaford, Delaware; Frederick Douglass PTA 52
Orlando, Florida; Progressive Firefighters Association 9
Columbus, Georgia; Exchange Club of Columbus 57
Rome, Georgia; Junior Service League of Rome, Inc. 57
West Point, Georgia; Professional Secretaries International 58
Coeur d'Alene, Idaho; Greater Kootenai County Fire Prevention Cooperative 59
Malad, Idaho; Malad Lions Club 60
Collinsville, Illinois; Madison County Firemen's Association 62
East St. Louis, Illinois; East St. Louis Volunteers 61
Peoria, Illinois; American Red Cross, Central Illinois Chapter 61
Presque Isle, Maine; Presque Isle Lions Club 72
South Portland, Maine; Thornton Heights Lions Club 72
Frederick, Maryland; Soroptomist International of Frederick County 74
Boston, Massachusetts; Justice Resource Institute, Inc. 75
Lucedale, Mississippi; George County Jaycees 77
Ballw-in, Missouri; West St. Louis County Chamber of Commerce 80
Cape Girardeau, Missouri; Zonta International, Cape Girardeau Club 80
Kansas City, Missouri; The Salvation Army 17
Pickering, Missouri; Pickering Lions Club 79
Albion, New York; Orleans/Genesee Rural Prevention Corporation, Inc. 19
Troy, New York; Greater Troy Chamber of Commerce 81
Beaufort, North Carolina; Carteret Community Action, Inc. 21
Allentown, Pennsylvania; The Burn Foundation (Elderly Program) 28
Aiken, South Carolina; Aiken County Independent Insurance Agents 89
Clemson, South Carolina; Blue Ridge Association of Insurance Women/Clemson Jaycees ... 89
Huron, South Dakota; Retired Senior Volunteer Program of Beadle County 32
Marion, South Dakota; Marion Area Jaycees/Marion Area Senior Citizens Center, Inc. ... 32
Nashville, Tennessee; Security Assistance For the Elderly, Inc. 95
Oak Ridge, Tennessee; Breakfast Rotary Club 95
Beaumont, Texas; American Red Cross, Beaumont Chapter 35
Galveston, Texas; Noon Optimist Club of Galveston 97
Post, Texas; Post Economic Development Corporation 35
TuUeta, Texas; Pettus Rotary Club 97

General Population-no specific age group (GP)

Special Polulation--handicapped individuals, ethnic groups, non-english speaking groups (SP)

APPENDIX 3

NCVFPP PROGRAM CATEGORIES

NCVFPP PROGRAM CATEGORIES

School-based Fire Prevention Programs (SB)

Elementary and Secondary School Fire Safety Education (SB:ES)

NCVFPP PROGRAM CATEGORIES

Preschool Fire Safety Education (SB:PS)

Residential Fire Prevention Programs (RB)

Home Fire Safety (RB:HF)

NCVFPP PROGRAM CATEGORIES

NCVFPP PROGRAM CATEGORIES

Residential Sprinklers (RB:RS)

Smoke Detector Installation and Maintenance (RB:SD)

NCVFPP PROGRAM CATEGORIES

Commercial Property Fire Prevention Programs (CP)

Population-Specific Fire Prevention Programs (PP)

Elderly Community Resident Fire Prevention and Survival (PP:EC)

Fire Prevention in Languages Other than English (PP:LG)

Special Focus Fire Prevention Program (SF)

Arson Prevention (SF:AR)

Wildland Fire Prevention (SF:WL)

Rural and Farm Fire Prevention (SF:RF)

Sioux Falls, South Dakota; McKennan Hospital Burn Program 93
L.oudoun County, Virginia; Loudoun Business and Professional Women's Club 39

Fire Prevention for the Handicapped (SF:HC)

Huntsville, Alabama; Exchange Club 45
Orlando, Florida; Progressive Firefighters Association 9
Coeur d'Alene, Idaho; Greater Kootenai County Fire Prevention Cooperative 59
New Castle, Indiana; American Red Cross, Henry County Chapter 63
Frederick, Maryland; Soroptomist International of Frederick County 74
Albion, New York; Orleans/Genesee Rural Prevention Corporation, Inc. 19
Rochester, New York; North East Area Development, Inc. 19
Tulsa, Oklahoma; American Red Cross, Tulsa Area Chapter 25
Allentown, Pennsylvania; The Burn Foundation (Mobility Impaired Program) 28
Altoona, Pennsylvania; Allegheny Township Fire Prevention, Inc. 27
Philadelphia, Pennsylvania; Associated Services for the Blind 27
Loudoun County, Virginia; Loudoun Business and Professional Women's Club 39
Keyser, West Virginia; Kiwanis Club of Keyser 105

APPENDIX 4

ORIGINAL MATERIALS PRODUCED

ORIGINAL MATERIALS PRODUCED

(listed by NCVFPP program category)

School-based Fire Prevention Programs (SB)

Elementan and Secondarv School Fire Safety Education (SB:ES)

Preschool Fire Safety Education (SB:PS)

Residential Fire Prevention Programs (RB)

Home Fire Safety (RB:HF)

ORIGINAL MATERIALS PRODUCED

(listed by NCVFPP program category)

Residential Sprinklers (RB:RS)

Smoke Detector Installation and Maintenance (RB:SD)

Commercial Property Fire Prevention Programs (CP)

Population-Specific Fire Prevention Programs (PP)

Elderly Community Resident Fire Prevention and Survival (PP:EC)

Fire Prevention in Languages Other than English (PP:LG)

ORIGINAL MATERIALS PRODUCED

(listed by NCVFPP program category)

Special Focus Fire Prevention Programs (SF)

Arson Prevention (SF:AR)

Wildland Fire Prevention (SF:WL)

Rural and Farm Fire Prevention (SF:RF)

Fire Prevention for the Handicapped (SF:HC)

*U.S.   G.P.O.:1993-719-591:80045

Federal Emergency Management Agency

United States Fire Administration

National Governors' Association

National Criminal Justice Association

For Additional Copies, please write to:
United States Fire Administration
Office of Fire Prevention and Arson Control
16825 South Seton Avenue
Emmitsburg, Maryland 21727

www.ingramcontent.com/pod-product-compliance
Lightning Source LLC
Chambersburg PA
CBHW081130170526
45165CB00008B/2627